Sanders Spencer

# Pigs

Breeds and Management

Sanders Spencer

**Pigs**
*Breeds and Management*

ISBN/EAN: 9783742837578

Manufactured in Europe, USA, Canada, Australia, Japa

Cover: Foto ©berggeist007 / pixelio.de

Manufactured and distributed by brebook publishing software
(www.brebook.com)

Sanders Spencer

# Pigs

# LIVE STOCK HANDBOOKS.

*Edited by* JAMES SINCLAIR, *Editor of* "*Live Stock Journal,*"
"*Agricultural Gazette,*" &c.

## No. V.

# PIGS.

## *BREEDS AND MANAGEMENT*

BY

## SANDERS SPENCER.

WITH A CHAPTER ON DISEASES OF THE PIG BY PROFESSOR
J. WORTLEY AXE, AND A CHAPTER ON BACON AND
HAM CURING BY L. M. DOUGLAS.

*ILLUSTRATED.*

London:

VINTON & COMPANY, LTD.,
9, NEW BRIDGE STREET, LUDGATE CIRCUS, E.C.

1897.

# CONTENTS.

# ILLUSTRATIONS.

---

# PIGS.

## BREEDS AND MANAGEMENT.

### CHAPTER I.

### BREEDS OF PIGS.

It has been freely asserted that the original wild pig, from which all our many cultivated breeds or varieties are descended, was of a rusty grey colour when young, the colour deepening as the pig reached maturity, and becoming a dark chestnut brown, with its hairs tinged with grey at the extremities as old age crept over it. For this opinion, which is expressed with great confidence by persons who have travelled considerably, there is much to be said. From residents in many foreign countries to which we have shipped pigs, we have learned that the semi-wild pig of the several countries is of a rusty or a slate colour, which with care in selection can be made of a lighter or of a darker shade. Climate and soil also undoubtedly affect the colour to a considerable extent; thus in Sierra Leone the pigs were described to us by the Hon. John Smith, one of the greatest benefactors of that country, as very small in size and bone, with comparatively little lean meat, and nearly black in colour, particularly the old pigs. In New Zealand the native pigs are very similar. In the colder portions of Siberia the pigs were stated by a resident

I

to be " very small, of a slate colour and *mostly bristles*." It
appears to be the same as regards the colour of the native
pigs of all countries—this has become fixed, or at all events
materially affected, by the colour of the lair in which it has to
make its resting and hiding place.  Thus in those countries
where the climate is such as to produce a profusion of dark-
coloured and rank herbage—as in tropical countries—there
the wild pig is found to be of a colour approaching black ; on
the other hand, in the temperate zone where the herbage is
sparse and the woodland partially free from undergrowth, the
colour of the wild pig is of a much lighter shade.  This may
have been brought about in two ways: the great heat in
the tropical countries may through many generations have
gradually had the same effect on the skins and even on the
hairs of the pigs as it has had on the hair and skin of human
beings ; whilst the mere fact that a pig when asleep or
resting was of the same colour as its surroundings, would
render it infinitely less subject to the attacks of its enemies,
either biped or quadruped, so that in the course of time the
predominating colour of the pigs would become lighter or
darker as the proportion of pigs which escaped destruction
were of the one or of the other shade of colour, and that
similar to the herbage or undergrowth of its environment.
It is impossible to estimate the proportionate effects of these
two causes, which undoubtedly affect the colour of the wild
pig, but we are inclined to think that the freedom from attack
and probable death enjoyed by pigs of a colour the nearest
approaching their lair is one of the chief agents.  On the
other hand, it appears to us that the climate of a country, and
its consequent variation in the quality, quantity and density
of the natural growth would have a most material effect alike
on the formation of the wild pig, on its disposition, and on its
natural aptitude to make meat quickly, or the reverse, and
having a greater or lesser proportion of fat.  For instance, in
a tropical country where the undergrowth is frequently a
tangled mass, free passage of the wild pig would be almost an

impossibility, whilst the ease with which the boar would be able to obtain a plentiful supply of bulky or fat-forming food would render it far less disposed, and indeed less able, to undertake those marauding or amatory excursions which would be certain to end in trials of strength and endurance.

The mere fact that these wild pigs had for many generations been bred from parents whose environments had rendered it unnecessary that any considerable exertion should be requisite on their part, or that the lords of the harem should be active and lithe, or of a quarrelsome nature, would in the course of time have the effect of completely altering the disposition and the form of the pigs. They would become less restless and consequently more inclined to grow fat instead of muscle ; those portions of the body used in the act of progression, in rooting for food, or in those fearful battles in which the marauding boars indulge, would become less developed, the bone of the whole frame would become lighter, and the hinder quarters and the depth of body would in course of time become proportionately greater. The pig itself would also naturally fine down, since that law, which by some persons has been called the survival of the fittest, would in their case have been to a great extent in abeyance. It is no doubt true that the lord of the harem would be the one boar physically the best endowed at the time, still there does not exist with these pigs the necessity for those powers of endurance, and natural defence so fully developed in wild pigs whose life is spent in colder countries, and where they have to go further afield to seek for their food, which mainly consists at certain times of the year of acorns, beechmast, chestnuts, &c. The colder climate would also render the pigs less lethargic, whilst the more solid and concentrated food would enable them to take longer journeys in search of food or companionship. The little pigs, too, would be more robust and of a more roaming disposition, so that they would, when young develop muscle, and make growth rather than meat. This variation in the mode of life and in the quantity and quality of the food, if continued for several generations, would most probably completely alter the character and the style of even a wild pig. If this be granted, then we need not seek further for the causes of those variations of type, colour, and feeding capabilities, which are noticeable in the pigs of different countries, and which have led some persons to imagine that the various breeds of pigs which are so dissimilar could not have originated from a common stock.

If the food, the mode of life, and the environment of the wild pig have had such a marked effect on its colour, its disposition, and its form, it can readily be understood that the building up of the many different breeds of the domesticated pig has not been so difficult a task as it appears to be at first sight, since in the effects of climate and food we have a far more powerful influence at work in the selection for breeding purposes of those pigs which possess, in a marked degree, the particular points which are the most sought after or prized by the owner. No stronger proof of the ease with which the form, and even many of the characteristics of the pig can be changed, has been afforded than in the marvellous changes which have taken place in the style and formation of the fashionable show-yard pig in the writer's time, or during the last forty or more years. The specimens now exhibited of some of the different breeds of pigs bear but the very slightest resemblance to those pigs of similar breeds which were honoured by the judges in the "fifties." Take, for example, the style, character, formation and colour of a Berkshire pig, which was considered to be of a correct type some forty years ago; it bears but the very slightest resemblance in colour, form and character to the Berkshire pig of the present day. Not only so, but have we not seen other changes almost as striking during the intervening period, and have not these changes been mainly brought about by fancy, fashion, or the effort to effect what the breeders imagined to be improvements?

LARGE WHITE SOW, ELPHICK'S DAISY.    Bred and Exhibited by Mr. E. Buss.

(Sire, Holywell Major; dam, Holywell Hop-picker II.)    Winner of First Prizes.

Our present object is not to attempt to show by what means these changes have been made—whether by selection alone, or by other means—but rather to furnish proof of the ease with which the form, character and qualities of a certain kind of pig may so greatly be changed, until the original is almost lost sight of, or so altered as to be scarcely recognised in the finished product of the period then current.

The Large White, or as it used to be termed the Large Yorkshire pig, also furnishes a striking instance of a similar transformation. In the time gone by—never, it is hoped, to return—a typical pig of the breed was considered to be one short in the snout, enormously heavy in the jowls, and thick in the shoulder, with back wide and fat, legs long, bone round and coarse, and a carcase of immense weight, mainly comprising lard, hide and bone. At the present time the Large White pig, or, as it is sometimes with truth called, the Improved Large Yorkshire, is an animal of a totally different type and character. It is true that a certain small section of the breeders of Large Whites are still vainly trying to stem the tide of progress by pinning their faith on, and giving their decisions in favour of helpless brutes, whose main, if not their only, claim to recognition is their size and inability to move with any amount of freedom, solely on the plea that a Large White pig must have size. In their mistaken zeal they entirely overlook the fact that a well-formed, compact pig, fine in bone, deep in carcase, and carrying much lean meat, will always weigh far better in proportion to apparent size than will one of the gaunt brutes they delight to honour, and not only so, but that the meat will be in infinitely better demand at much higher prices.

If it were advisable, we could mention other breeds of pigs which have been pretty well *improved* off the face of the earth, the show-yard decisions, based on little, if aught else than fancy points, being the chief cause of the disaster. Then on the other side there is a fatal risk. Within the last fifteen years a breed of pigs was boomed because it was alleged that the pigs of the breed had not been improved. We have, in this, an example furnished of the other extreme, and one also which proved unfortunate, since these so-called unimproved pigs really possessed many of the qualities of unimproved pigs, and the claims made for them were too literally correct.

One hears it occasionally stated that fashion in food has a determining influence on the form and qualities of the pig of a district or county; but we are inclined to the opinion that prejudice and local conditions are both of them equally potent factors. We would ask by what other name than prejudice, can be called the utter folly of the residents in certain districts who persist in " boycotting " the carcase of a pig desirable in every other way, except that its skin, before slaughter, was white or the reverse. That this most childish course of action is very common in the less enlightened districts is an admitted fact, and what is still more strange, the very pigs which are honoured solely on account of the colour of their hair and skin, often furnish pork which is of an inferior character compared with that which ranks highest in favour with the general public.

The United States furnish us with what we consider, if not an unanswerable, at least a very strong argument in favour of the contention that local conditions have exerted a potent influence on the form and character of the pig, well nigh during the whole period in which the American breeders of pigs have been attempting to originate so-called new varieties of pigs, or in cultivating those English breeds of pigs of which they have imported very large numbers. One of the main objects for which pigs have been kept in their millions, has been to convert the immense crops of Indian corn or maize into marketable products, or as the Americans have it, so that the maize should walk to market. As every one is well aware, maize consumed by hogs results in very fat pork. Fat pork lends itself better to the packers' trade than does lean pork, for one reason that it does not become so salt, whilst it is also stated that it will keep better under certain conditions. A large proportion of the salted pork consumed in the States has been eaten under climatic and other conditions which rendered fat salted pork more suitable and necessary than lean salted pork. The value of lard in European countries was such that the fat pork could be rendered or melted down and exported to this and other countries at a profit, whilst the conditions of life amongst the poorer classes in England, Germany, Sweden, and other European countries were such that the low-priced salted pork from America was freely bought and consumed in preference to the leaner and much more expensive home-grown pork. No wonder, then, that under these and other conditions exerting a similar influence, the pig breeders in the States favoured those pigs which were considered to be the best lard and fat-meat-making machines. The principal points sought for in pigs in America were a wide and fat back, a thick neck, and a heavy jowl, since a pig possessing these points was certain to be a good maize-into-lard converting machine. Then still another influence was at work; it was found that the smaller kinds of pigs were first-rate fat producers, but that they either became fat at too young an age, or did not produce pork of a satisfying or wearing nature. What was wanted by those who fed the thousands of men who spent the long and cold winters in the back woods, was pork, of which a small quantity satisfied the men's wants, or in other words, meat from much older pigs. This influence was also aided by the general craze of those American farmers, in whose life there was little opportunity to favour competition with their neighbours, or comparatively no other amusement or employment than hard work, to make their fat pigs the medium of speculation or gambling. Sweepstakes were started for the heaviest pigs in the various districts. When the weighing-in day arrived—you could scarcely call it judging—the majority of the farmers within many miles of the place of meeting, which was generally

the nearest town, would attend, and a real good day of it
was made. The pride of place for the season amongst hog-
raisers was not won without a considerable outlay of food and
attention spread over a year or two at least, as to have any
chance of success the fat hog must have scaled at least
1,000 lbs., whilst 1,200 to 1,300 lbs. hogs are recorded.
Most of us resident in England are unable to form an idea
of what these fatted monsters must be like, since it is most
unusual to see a pig in this country weighing more than
1,000 lbs. That these immense pigs must have had coarse
heavy bone to carry the enormous weight is evident, and
further, the proportion of fat in the carcase must have been
very large, whilst the cost of producing it would be out of all
proportion to its value as meat.

From the above remarks it is evident that the style, quality
and formation of the pig have been greatly influenced in the
States by conditions which do not at the present time exist
in that country; one cannot, therefore, be surprised when
the opinion is expressed that the monetary affairs and the
mode of life of the inhabitants, combined with the climate and
the natural produce of the soil, have an almost preponderating
influence on the kind of hog bred in a certain country. So
far as one can learn, the major portion of the English pigs
imported into America some thirty or forty years since were
of the smaller and fatter type of early maturing pigs, a short
snout, an inordinately heavy jowl, thick neck, and fat back,
being the chief points sought for and valued. Pigs of
this character were fashionable at the time in this country,
prizes were awarded to them, and owing (fortunately for
this country) to their comparatively small numbers, the
cost of them to our American cousins was high, and this,
too, added to their popularity amongst a people who are
said not to reckon the cost if it be but possible to secure
something rare and fashionable. For the mere winning of
prizes at the State fairs or shows, and the producing of lard,
these imported bladders of animated lard were so great a

success that as many as possible of them were for a time purchased at high prices by those who made large profits in buying English stock, taking them over the Atlantic and selling them to the happy possessors of a good supply of almighty dollars.

A commercial influence, however, soon made itself felt in the States. The ordinary, apart from the fancy hog raiser, required a pig of more growth and of far greater weight when it arrived at maturity. Crosses with the small fat pigs were made with the larger and coarser pigs, such as those called Cheshire and Chester whites, Suffolk whites, &c., with the result that several so-called new varieties, such as the Poland China, the Victorias and others were produced, and so well did the compound pig, now christened the Poland China, take on, that a large majority of the many millions of pigs in the States are said to be of this breed or of a similar character to it. It appears to be impossible to learn just what breeds of pigs were most used in the manufacture of this Poland China, of which the breeders in the States are said to be so proud as to lead them to assert that it "licks all hog creation"; but so far as an outsider can judge it would appear as though a very similar course of proceeding was followed as that adopted by our own breeders, by whose assistance the fashionable Berkshire was produced, except that in the case of the Poland China some kind of pig having a bent or broken ear, or one apparently broken, was employed, whilst in the Berkshire a prick-eared pig was used. There are apparently two other differences; owing to the influence of the Neapolitan pig the Berkshire has a thinner skin, of a lighter shade, and is less coarse altogether than the Poland China, which was apparently built up without any, or only a little, aid from the delicate Neapolitan. With these exceptions the breeders of the Poland China may fairly claim to have produced a breed of pigs possessing most of the valuable points—even to the white markings—of our modern Berkshire. Then, again, take the so-called

Victoria, a white pig of a form and appearance very similar to our Middle White, to which it no doubt owes a great proportion of its good qualities. It is, of course, impossible to know, except from the originator of a so-called new variety, as to the precise crosses used, but we can give a pretty good guess, or perhaps it would be more judicious to remark that by crossing a Berkshire sow and the resultant produce for three or four generations with pure-bred Middle White boars, a new variety of pig almost exactly similar to the Victorias would result. It might be urged that the Middle White pig was not known in the States some twenty or thirty years since, and this contention would be partially sound, but the sting would be extracted from it were the addition of the following words, *under its own name*, made to the assertion. Middle White pigs were frequently shipped to the States where they blossomed forth as either Yorkshires or Suffolks; the smaller pigs as a rule did duty for Yorks, and the larger and more growthy specimens became Suffolks. Boars of the last kind, if used for a few generations on a Berkshire sow, would result in an American Victoria, or of a pig so like it that it would be difficult to discover any difference. Indeed, some of those jealous breeders of other varieties of pig are reported to have said that attempts at least have been recently made to improve even the Victoria, by the infusion of a modicum of Middle White Yorkshire blood. Another variety of pig which our American cousins claim to have produced is the Duroc-Jersey, but here, again, we should imagine that the step taken by the originators of this kind of pig has been very similar to those which have been adopted by other breeders of pigs.

A decided and complete revulsion of feeling is certainly taking place in the States, the very large and coarse pigs are doomed, and the same sad fate as surely awaits those pretty little round tubs of animated lard.

The causes for this are many, but the gradual change in the mode of life of Americans generally, the very considerable increase in the proportion of the inhabitants of the United States who now reside in towns, the requirements of the export trade—which have totally altered during the last quarter of a century—and the use of cotton seed oil most extensively as an adulterant of lard, are a few of the chief influences. The breeders of pigs in the States will have great difficulty in conforming to the wants of the market, since the pedigree or pure-bred craze has a deep hold on our American cousins, whilst the colour fancy or prejudice is by no means unknown amongst pig-keepers. Then again the principal pig food is corn, or, as we call it, maize; this will have a certain influence in counteracting the efforts of the exporters of hog products to the British Isles, as well as of those packers who specially study the town or city markets in the various States. It is not within our province to point out the means by which the indispensable alteration will have to be made; this we can leave with every confidence to the pig-breeders in the States. They will soon discover if they possess, in any one of their breeds or crosses of pigs, animals which will grow quickly and produce at six or seven months a long, deep, lean, and light fore-quartered carcase of pork, weighing some 160 lbs. It may be doubtful if this kind of pig is at all general in the States; if not, it will be in the near future, or as soon as the pig-breeders grasp the somewhat novel idea.

So long as we were discussing the manufacture or the building up of the different varieties of pigs in the United States, we were on tolerably safe ground, since nearly all the facts related have occurred either within our own memory or since the establishment on the American continent of several of those stock papers which sprang into existence long before similar periodicals were general in this country. Not so with regard to the history of those breeds of pigs which have been considered with us for many years as pure-bred. It is true that some writers on pigs have given the names of various persons as having originated certain breeds, but a thorough examination of the foundations for these historical statements results in the discovery of a considerable amount of assumption. If we take the White, or, as it is called, the Yorkshire breed, we find that different writers apportion the amount of credit to a number of persons very variously. Some at least of the chroniclers appear to have selected for special honour those persons from whom they may have bought those pigs which proved most successful in their hands, or it may be that certain successful strains of pigs of which they may have become possessed could be traced back to a certain breeder, who was then written up as one to whom the breeders of the particular kind of pig were most largely indebted. These we find duly written of as one of the first improvers of the Yorkshire. Then another writer will declare, with equal confidence, and perhaps the same amount of truth, that to Wainman belongs the chief credit; other names, such as Wiley, Nutt, Jolley, &c., will occur to those of our readers who have studied the pig literature of recent date. Personally, we frankly confess that we possess little, if any, knowledge of the first improvers of our present breeds of pigs. It is true that we might mention a number of names of persons who during the last forty years have been amongst the most successful exhibitors of pigs, and who might by some persons be considered to have thus made good their claim to be enrolled as the improvers, if not the originators, of our various breeds of pigs; but here a difficulty would arise as to at least some kinds of pigs, since it is a well-known fact that many of the chief exhibitors used to purchase the major portion of the pigs which they showed successfully. This has been the case more particularly with White pigs, as these have been generally kept in those northern counties of England where large numbers of intelligent men, having earned high wages in the following of their occupations in the various mills, factories, &c., sought for some pursuit which promised to afford a certain amount of excitement, pleasure, and profit. Nothing so completely lent itself to the satisfying of this desire as the breeding of live animals—dogs, rabbits, pigeons, poultry and pigs. The last-named, giving the most promise of profit,

Holywell Dublin.    Pride of Oxford.    Holywell Count.

LARGE, MIDDLE, AND SMALL WHITE PIGS.

Winners of numerous First and Champion Prizes.    Bred and exhibited by Mr. Sanders Spencer.

perhaps, had the greatest number of votaries. A Yorkshire-man appears to have an inherent love for a horse or a pig; therefore it is not surprising to find that the pigs kept in many portions of the counties of Lancaster and York some thirty years since, were equal, if not superior, to those found in any other part of the country. Those numerous district shows, which are frequently held on the last day of the week, may have been the outcome of the fondness of the residents for live animals, but it appears equally as probable that the con-tinuance of the shows resulted in an increased amount of attention being paid to the breeding of the various kinds of stock which possessed fancy points. Some thirty or forty years ago, it was nothing unusual to find the display of pigs at the best of these local shows well-nigh equal in merit and numbers, particularly of the White breeds, to the displays of pigs at the largest shows in the southern counties. Unfortu-nately, the classification usual at these district shows was not such as tended to purity of breeding, the prizes very fre-quently being offered for the best pigs of any White breed. This led to the breeders paying their chief attention to the production of the biggest and fattest pigs possible, irrespective of those points which are supposed to be the distinguishing characteristic of the sub-varieties into which the Yorkshire pigs have for some years been divided. The breeders of these pigs were not slow in discovering that the easiest way of obtaining early maturity—a most important point, as the classes for young animals were very numerous—was by cross-ing the Large, Middle and Small White pigs almost indis-criminately, although perhaps the most favourite plan was to mate a Small White boar with a Large White sow possessing as much substance and quality as possible; if there happened to be a strain of Middle White blood in the sow no objection would be made. The resultant litter would, of course, vary somewhat in character and size, but nearly all the youngsters would partake of the early maturing qualities of the sire, and, to a varying extent, of the quick growth of the dam. This would enable the owner to bring out some heavy-fleshed and fine quality boars and yelts for the classes limited to pigs under six, nine, or twelve months old, classes which were looked upon with special favour by the exhibitors at the district summer shows, because representatives of those ex-hibitors who used to have large teams of pigs on the show circuit generally either attended these small shows to pur-chase some of the best of the animals exhibited or had agents for that purpose. The successful exhibition of these young boars also served as a good advertisement for them, should they remain unsold and be retained by their owners for service.

Not only was this crossing of the three sub-varieties of Yorkshire generally adopted by a considerable proportion of the pig-breeders in certain districts, but in years gone by even Berkshire sows are said to have been mated with Large White boars, and the largest of the boar pigs, and those taking most after the sire in form, size, and character, have been successfully exhibited at our largest shows in the classes for Large White boars. As many of the exhibition pigs of the fifties or sixties, and even a later decade, shown in the classes for Large, Middle, and Small White breeds were said to be the result of this promiscuous breeding, it is almost impossible to give any historical record of value as to the early improvers of pure-bred White pigs. It is well known that several Yorkshire-men who bred a few pigs also spent a considerable portion of their spare time in hunting up pigs which either had won prizes at the local shows or which appeared likely to grow into animals good enough to prove successful at the larger shows. These pigs then passed into the hands of those who would exhibit them at the Royal and others of the best shows without any pedigree given, and very frequently without the slightest information as to the name of the breeder, who, more often than not, was unknown.

Amongst the early successful breeders and exhibitors of Yorkshire pigs have been Tuley, Wainman, Matthew Walker, Lieut.-Col. Cooke, R. E. Duckering, Earl of Radnor, Earl of

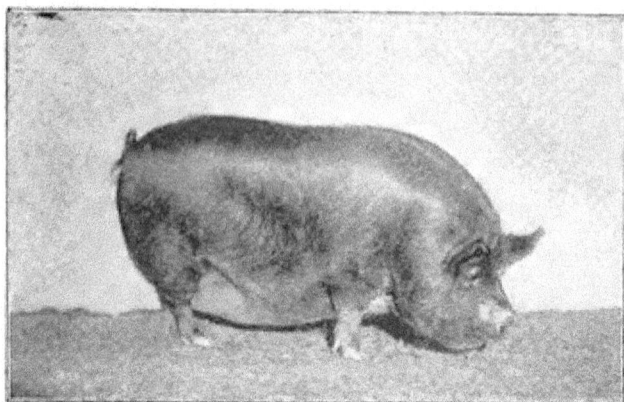

SUFFOLK SOW.

Winner of First and Champion Prizes.

Bred by and the property of Mr. J. A. Smith, Akenham, Ipswich.

Boar, 1 Year old, 'Royal. Warwick. 1st

BERKSHIRE BOAR.

Exhibited by Sir Humphrey F. de Trafford, Bart.

Ellesmere, J. and F. Howard, Peter Eden, T. Strickland, the author of this book, and others; whilst later came Philip Ascroft, C. E. Duckering, F. A. W. Jones, John Barron, T. Collinson, A. C. Twentyman, Edwin Buss, D. R. Daybell, Denston Gibson, Joseph Ashforth, N. J. Hine, Frank Allmand, Sir Gilbert Greenall, Bart., C. Ecob, &c.

In connection with the Small Black or Suffolk, or again, as it is styled in the States, the Essex, who has not heard or read of the successful efforts of Lord Western, Fisher Hobbs, Stearn, Crisp, and in later years of Sexton, Smith, Petitt, and the late Duke of Hamilton? There is but little doubt that the above-named and a few other breeders brought the Small Black pig within measurable distance of perfection as a converter of corn into meat—of a certain kind.

With the Berkshire, as with the Small Black, the majority of the exhibitors have been content to show only those animals which they have bred. One is thus much more easily enabled to point out those breeders of Berkshires who have been successful in producing the best animals of the fashionable type of the Berkshire pig, an animal which was particularly fortunate in securing for its admirers many of the most successful breeders of cattle and sheep of some thirty or forty years ago—men who, above everything, were practical and observant. The beneficent effects arising from this breed being taken in hand by thoroughly business men rather than by mere fanciers were soon apparent. Anyone who was able to admire the best specimens of a variety of stock other than the particular breeds which he was at the time breeding could not have failed to admire the Berkshire pig of some twenty-five years since. It possessed in a marked degree those qualities so highly prized at the present time; it was both handsome and robust, of good size and length, and yet free from coarseness, and was fairly prolific, whilst the little pigs were hardy, and made a sufficiency of growth to furnish a lean, thick-fleshed carcase of pork by the time they had reached a year old. This was at a period anterior

to the colour and quality craze, a craze which has tended to
reduce the commercial value of many of our varieties of stock.
There is but little doubt that the Berkshire pig, in common
with the Small White and Small Black pigs, suffered considerably, at a later period, at the hands of the mere fancier
and showman, whose first thought was too frequently the
production of prize-winners of such a type and formation as
was not common to the breed, nor very frequently the most
serviceable.  Then the selection of only such judges as
favour the fashionable type has usually ended in practical
men turning their attention to the rearing of animals of a
less fashionable but of a far more generally useful variety.  It
is also asserted, with a considerable amount of confidence,
that the old-fashioned useful Berkshires suffered very much
from our American cousins taking them up so keenly, since
the majority of the purchasers from the States were either
men of almost unlimited means, who had taken up the breeding
of pedigree stock as a mere hobby, or for personal purposes ;
or else the buyers were agents of these capitalists, and in
a few cases speculators who answered to a class which in
England is known as dealers.  The first and chief aim of
these importers was to purchase prize-winning animals, or
such as conformed to their fancy ideal.  A few of the points
in a pig most sought after by these buyers were a short, turned-
up snout, a heavy jowl, thick neck, wide shoulders, and a fat
back.  A number of our most successful breeders of Berkshires
naturally bred their pigs of the form required in order to
share in the very lucrative shipping trade, and the result has
not proved to be permanently beneficial to the Berkshire pig,
which is declared by Poland China breeders to be not now so
much in demand in the States for general purposes as it was
some fifteen or twenty years ago.  Amongst the most successful breeders of Berkshires, some twenty or thirty years
since, must be mentioned Mr. Joseph Smith, Captain Stewart,
Mr. Russell Swanwick, Mr. John Lynn, Mr. Heber Humfrey,
Mr. Barnes, of Solesbridge, Mr. Pittman King, Mr. Richard

BERKSHIRE SOW, HER MAJESTY. Bred and exhibited by Mr. Nathaniel Benjafield.

Winner of Five Champion and other Prizes.

Fowler, Mr. Henry Tait, of the Royal Farms, the Hewers, and many others whose names are equally deserving of notice, but which do not at the time of writing recur to us. Of later years, Mr. Edney Hayter, Mr. James Lawrence, Mr. A. S. Gibson, Mr. J. W. Kimber, Mr. A. Hiscock, Mr. W. A. Barnes, of Birmingham, Mr. N. Benjafield, Mr. A. Darby, Mr. H. Lambert ; and, still more recently, the Duke of York, Prince Christian, the Earl of Carnarvon, Sir Humphrey de Trafford, Sir James Blyth, Mr. J. A. Caird, Mr. H. P. Green, Mr. Edwin Buss, Mr. P. L. Mills, Mr. Adeane, and many others have kept up the fame of the Berkshire.

The history is brief of one of the oldest of our varieties of pigs, the mahogany or grizzled pig, which has acquired the name of the Tamworth. This very hardy, if not handsome, pig was extensively bred in Leicestershire, Staffordshire, and Northamptonshire, and one or two adjoining counties, early in the present century. Even before the battle of Waterloo was fought, these dark-red and grisly pigs were general in those portions of the Midland counties where considerable numbers of oak and beech trees were grown ; large droves of these pigs were sent into the woods and forests, where they spent the chief part of the autumn and early winter, finding the major part, if not the whole, of their food. In later times, after the country had become enclosed, and a considerable portion of the woodlands was converted into arable and then into pasture land, farmers found that a pig of a somewhat quieter disposition, and one which would fatten more readily, was necessary. This change was accomplished in many cases by crossing the long-snouted, prick-eared, sandy and grey with black spots pigs with pigs having a strong infusion of Nea-politan blood. Many persons also used the white pig, which was common in Bakewell's time, and which he, mainly by in-breeding and selection, rendered both delicate and more easily fattened, and then termed the white Berkshire breed. The result of this mixture was the plum-pudding or the black, white, and sandy pig, to which reference is made in

another portion of this work. In certain districts of Staffordshire and the adjoining counties, the breeders of these
mahogany-coloured pigs took considerable pains by selection
to increase the feeding properties of their pigs without losing
their distinctive colour. These pigs were not particularly
quick feeders, but they were prolific, and when well fattened,
furnished a splendid carcase of pork, nicely intermixed with
lean.

Some eighteen or twenty years since, when our bacon
curers opened the campaign against the then fashionable
short, fat, and heavy forequartered pig, which carried nearly
two-thirds of its weight in the least valuable part of the carcase, the Tamworth was taken up by the late Mr. G. M.
Allender and a few others as the type of pig to cross with
the fat pigs, and so render them of more value to the bacon
curer. The introduction into the prize list of the Royal
Agricultural Society of a separate class for Middle White
Yorkshires, which had hitherto been shown in the any other
breed classes, gave the breeders of Tamworths a most favourable opportunity to bring their favourites before the public.
For some years the Tamworths have had separate classes at
the Royal shows, with the result that many supporters of
the breed have sent their pigs for exhibition. As no particular exhibitor appeared to be so very much more successful than his competitors, and as the majority of them
adopted the system of showing only those pigs bred by
themselves—a system which is productive of far greater
improvement in a variety of stock than when a few wealthy
men buy up all the best of the pigs of each season, and
thus kill competition—large and very good entries of Tamworths were seen, particularly at the shows held at Birmingham in November. Amongst those who, besides the
late Mr. G. M. Allender and the Aylesbury Dairy Company, of which he was manager, have successfully exhibited
Tamworths might be mentioned Lord Auckland, Messrs.
Egbert de Hamel, J. A. Herbert, J. Hill, R. Ibbotson,

TAMWORTH BOAR.

Winner of First Prizes.    Bred and exhibited by the Aylesbury Dairy Company.

W. H. Mitchell, John and Joseph Norman, W. D. Philip, J. R. Randall, R. N. Sutton-Nelthorpe, T. Clayton, T. Tompson, G. T. Whitfield, Boston, J. Dunnington Jefferson, Howard, Howard Taylor, J. A. Herbert, J. H. Jordan, R. Boddington, &c.

The attempt to draw up a standard of value of the various points a perfect pig of a breed should possess is at the best a difficult one, whilst the certainty that objections will be found to it does not render the task less easy. With the exception of the scale of points for the Berkshire we have in the following drawn only upon our own opinions and our rather extensive experience gained during the last thirty-five years. The Berkshire scale is a company affair. In drawing up the one which follows we availed ourselves of the valuable advice of some of the most careful and the oldest of the breeders of the variety. We may remark in passing that a scale of points for Berkshires which we compiled three or four years since, at the request of some colonials, varied but little from the one now given, the main difference being in the proportion of points awarded to colour and to portions of the forequarters, on which a lesser value was set. For this a good reason exists, as the Society in whose special charge rests the interests of the breeders of the Berkshire pig have determined that certain variations in the markings or the apportionment of black and white shall disqualify the pig possessing them as a winner of a prize at a show. Just whether this will prove of benefit or the reverse to the breed is not now pertinent, but should the prognostications of those who are not extreme purists prove correct, the Berkshire will not be alone in having its usefulness impaired by those who appear to value more highly the fancy points than those of a more practical nature.

In the British Isles we have a number of breeds of pigs which are more or less local in character, but at the present time only six of these are acknowledged as pure breeds and have registers to record their pedigrees. They are the Berkshire, the Tamworth, the Small Black and the Yorkshire, subdivided into the Large, the Middle and the Small White breeds.

We were rather inclined to shy at the attempt to give our views as to the source or origin, and the good or indifferent qualities of the varying breeds, since amongst the breeders of each variety will be found those who will not hesitate to give expression to their opinion that their own particular favourites are harshly criticised or insufficiently belauded. An honest attempt has been made to satisfy all legitimate requirements, and there the matter ends so far as we are concerned. A " standard of excellence " has been drawn up by certain breeders, but they have fallen into the common and great error of imagining that the mere enumerating of certain practical and fancy points is sufficient without making any attempt to apportion the value or importance of these several points. This want appears to the author to render the standard of less value, particularly to the young breeder, who may be doing his level best to provide his pigs with a tail " stout and long, but not coarse, with tassel of fine hair," under the mistaken idea that each one of the points of perfection counts for as much as the other at the hands of the fancy judge ; whereas, with a scale of the value of each of the desirable points, the tyro can see at a glance which are those points for which he should more particularly seek when purchasing his foundation stock, in order that the pigs which he may breed may possess those commercial points that are of infinitely more importance to him and the country at large than those imaginary perfections, such as a tassel of hair on the long, thin, but not coarse, tail.

### SCALES OF POINTS.

#### BERKSHIRES.

*Colour.*—Black, with white blaze on the face, white feet
    and white tip to the tail   ...   ...   ... 10
*Head.*—Long and light, wide between the ears   ... 4

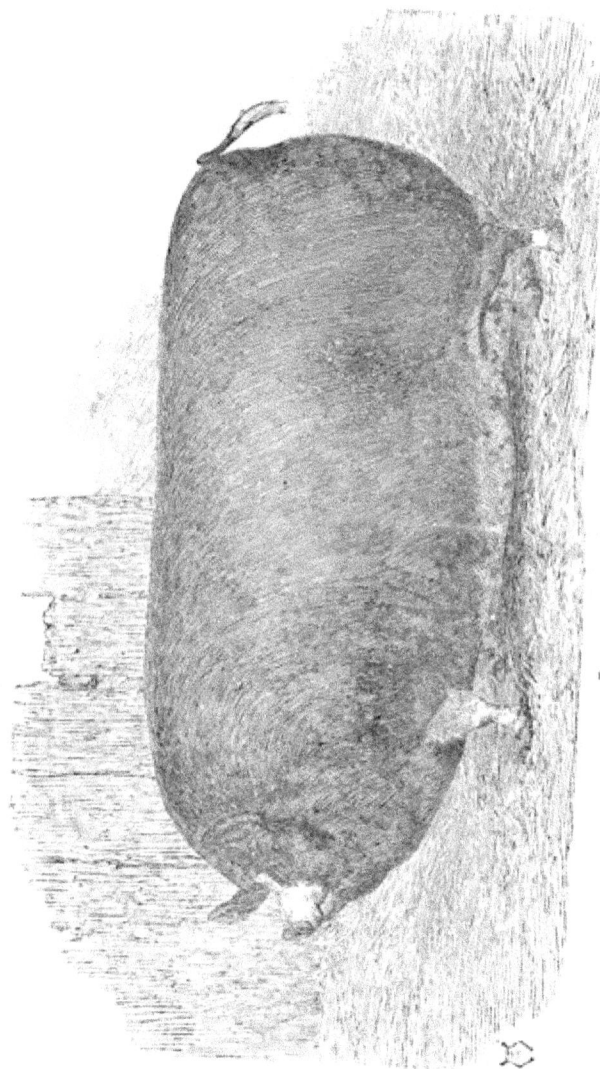

BERKSHIRE SOW, HIGHCLERE.

Winner of First and Champion Prize at R.A.S.E. Show, Windsor, 1889.     Bred and exhibited by Mr. T. H. E. Hayter.

## BERKSHIRES—*(continued)*.

*Ears.*—Thin, pricked and fringed with fine hair ... 3
*Jowl.*—Small and light ... ... ... ... 2
*Neck.*—Long and muscular ... ... ... ... 3
*Chest.*—Wide and well let down ... ... ... 5
*Shoulders.*—Oblique and level on top ... ... ... 4
*Girth.*—Around the heart ... ... ... ... 4
*Back.*—Long and straight ... ... ... ... 5
*Sides.*—Deep ... ... ... ... ... 5
*Ribs.*—Well sprung ... ... ... ... 4
*Loin.*—Strong, not drooping ... ... ... 2
*Belly.*—Full and thick, with at least twelve teats ... 3
*Flank.*—Thick and well let down ... ... ... 2
*Quarters.*—Long and wide from hip to tail ... ... 6
*Hams.*—Broad, full and meaty to the hocks ... ... 8
*Tail.*—Set high, not coarse ... ... ... ... 2
*Legs.*—Straight and with flinty flat bone ... ... 5
*Ankles.*—Strong and compact ... ... ... 3
*Pasterns.*—Short and yet springy ... ... ... 3
*Feet.*—Firm and strong ... ... ... ... 2
*Evenness.*—Freedom from wrinkles on skin ... ... 4
*Coat.*—Fine and long ... ... ... ... 3
*Action.* ... ... ... ... ... ... 4
*Symmetry.*—General style and contour showing evidence
  of careful breeding ... ... ... ... 4
                                                    ———
                                                    100

### Objections.

Narrow forehead or short pug nose.
*Ears.*—Thick, coarse, white or much inclined forward.
*Jowl.*—Fat and full.
*Neck.*—Short and very thick and fat.
*Chest.*—Narrow, with both forelegs apparently coming from
  almost the same point.
*Shoulders.*—Coarse, heavy, or wide and open on the top.
*Girth.*—Light round the heart and foreflank light.
*Back.*—Weak and hollow when pig is standing at rest.

### BERKSHIRES—*(Objections continued).*

*Sides.*—Shallow, not well let down between the forelegs.

*Ribs.*—Flat or short curved ; light back rib.

*Loin.*—Narrow and weak.

*Belly.*—Flaccid or wanting in muscle, or gutty, or podgy.

*Flank.*—Thin, and not well let down.

*Quarters.*—Short, narrow or drooping.

*Hams.*—Narrow, wanting in depth, or deficiency of muscle in second thigh.

*Tail.*—Coarse and set on low.

*Legs.*—Crooked, weak, and with round and coarse bone.

*Ankles,*—Extra large, round, and weak.

*Feet.*—Flat, splayed and extra wide or large.

*Evenness.*—Wrinkles on sides, neck, or shoulders.

*Coat.*—Coarse, curly, bristly, or maney, with fringe along top of neck and shoulders.

*Action.*—Sluggish and clumsy.

*Symmetry.*—Predominance of certain points, especially heavy shoulders or forequarters generally, with weak loin and light hams.

#### *Disqualifications.*

*Colour.*—White or rusty patches of hair.

*Boars.*—Rupture ; one testicle only down.

*Sows.*—Deficiency in or very irregularly placed or blind teats; injured or diseased udder.

### SMALL BLACK.

Similar to Small White, except in colour, which should be black without white.

### TAMWORTH.

| | | | | |
|---|---|---|---|---|
| *Colour.*—Golden red without black spots | ... | | ... | 5 |
| *Head.*—Long, snout straight, wide between the ears | | ... | | 4 |
| *Ears.*—Thin, pricked and fringed with fine hair | | | ... | 3 |
| *Jowl.*—Small and light | ... | ... | ... | ... | 2 |
| *Neck.*—Long and muscular ... | ... | ... | ... | 3 |

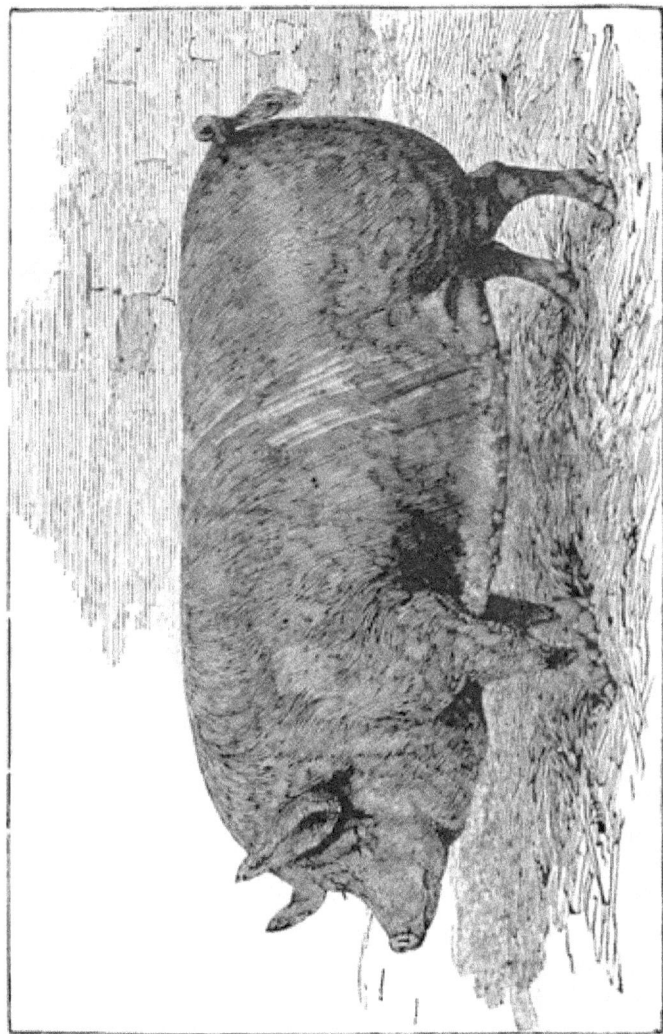

TAMWORTH SOW, PLYMOUTH QUEEN.

Winner of First Prizes at R.A.S.E. and other Shows.    Bred and exhibited by Mr. John Norman, jun.

## TAMWORTHS—(continued).

| | |
|---|---|
| *Chest.*—Wide and well let down ... ... ... | 5 |
| *Shoulders.*—Oblique and narrow on top ... ... | 4 |
| *Girth.*—Around the heart ... ... ... ... | 4 |
| *Sides.*—Deep ... ... ... ... ... | 5 |
| *Ribs.*—Well sprung ... ... ... ... | 5 |
| *Loin.*—Wide and strong, not drooping ... ... | 4 |
| *Belly.*—Full and thick, with straight underline and at least twelve teats ... ... ... ... | 2 |
| *Flank.*—Thick and well let down ... ... ... | 4 |
| *Quarters.*—Long, wide and straight from hip to tail ... | 7 |
| *Hams.*—Broad, full and meaty to the hocks ... ... | 8 |
| *Tail.*—Set on high, not coarse ... ... ... | 3 |
| *Legs.*—Straight and with flinty flat bone ... ... | 6 |
| *Ankles.*—Strong and compact ... ... ... | 4 |
| *Pasterns.*—Short and yet springy ... ... ... | 2 |
| *Feet.*—Firm and strong, not splayed ... ... ... | 3 |
| *Evenness.*—Freedom from wrinkles in skin ... ... | 2 |
| *Coat.*—Long, straight and silky ... ... ... | 3 |
| *Action.*—Free and clean ... ... ... ... | 3 |
| *Symmetry.*—General style and contour giving evidence of good breeding ... ... ... ... ... | 4 |
| | 100 |

### Objections.

*Head.*—Narrow forehead or upturned nose.
*Ears.*—Thick and coarse, or inclined forward.
*Jowl.*—Thick and coarse, similar to Berkshires.
*Ribs.*—Flat, or short curved ; light back ribs.
*Loin.*—Narrow or weak.
*Belly.*—Flaccid or wanting in muscle, gutty or podgy.

### Disqualifications.

*Colour.*—Black hairs or black patches on the skin.
*Boars.*—Same as for Berkshires.
*Sows.*—Same as for Berkshires.

## Large Whites.

| | |
|---|---|
| *Colour.*—White, freedom from blue spots on skin desirable | 2 |
| *Head.*—Long and light, wide between the ears | 4 |
| *Ears.*—Thin, long, slightly inclined forward and fringed with fine hair ... | 3 |
| *Jowl.*—Small and light | 2 |
| *Neck.*—Long and muscular ... | 3 |
| *Chest.*—Wide and well let down | 5 |
| *Shoulders.*—Oblique and narrow on top | 4 |
| *Girth.*—Around the heart | 4 |
| *Back.*—Long and straight | 5 |
| *Sides.*—Deep | 5 |
| *Ribs.*—Well sprung | 5 |
| *Loin.*—Broad and not drooping | 3 |
| *Belly.*—Full and thick with at least twelve teats | 2 |
| *Flank.*—Thick and well let down | 4 |
| *Quarters.*—Long, wide and straight from hip to tail | 7 |
| *Hams.*—Broad, full and meaty to the hocks | 8 |
| *Tail.*—Set on high, not coarse | 3 |
| *Legs.*—Straight and with flinty flat bone | 6 |
| *Ankles.*—Strong and compact | 4 |
| *Pasterns.*—Short and yet springy | 2 |
| *Feet.*—Firm and strong | 3 |
| *Evenness.*—Freedom from wrinkles on skin | 2 |
| *Coat.*—Long, straight and silky | 4 |
| *Action.*—Free, clean and not rolling in hind quarters | 5 |
| *Symmetry.*—General style and contour showing evidence of careful breeding | 5 |
| | 100 |

*Objections.*—Same as for Berkshires.

### Disqualifications.

*Colour.*—Black hairs or black spots.
*Boars and Sows.*—Same as for Berkshires.

LARGE WHITE BOAR, HOLYWELL, WINDSOR.

Winner of First and Champion Prizes at Leading Shows.     Bred and exhibited by Mr. Sanders Spencer.

## MIDDLE WHITES.

*Colour.*—White, freedom from blue spots on skin desirable    2
*Head.*—Short and light, wide between the ears    ...    4
*Ears.*—Thin, small and pricked, with fringe of fine hair...    3
*Jowl.*—Small and light    ...    ...    ...    ...    2
*Shoulders.*—Oblique and narrow on top    ...    ...    4

Others same as for Large Whites.

*Objections.*—Same as for Large Whites.

*Disqualifications.*—Same as for Large Whites. .

## SMALL WHITES.

*Colour.*—White    ...    ...    ...    ...    ...    2
*Head.*—Short, wide between the ears...    ...    ...    4
*Ears.*—Thin, small and pricked, with fringe of fine hair...    3

Others same as Large Whites.

*Objections.*—Same as Large Whites.

### *Disqualifications.*

*Colour.*—Black hairs or black or blue spots on skin.
*Boars and Sows.*—Same as for Berkshires.

Besides the foregoing six varieties or sub-varieties—which are now acknowledged as distinct, since the pedigrees of each are recorded by the National Pig Breeders' Association, whilst the Berkshire has also the British Berkshire Society to specially look after the interests of its breeders—there are many local breeds more or less distinct and which have for a great many years been cultivated within areas of varying size. The establishment of bacon factories, the spread of knowledge as to the most generally suitable style and formation of a carcase of pork and the increased facilities of transport, have all helped to reduce the number of the specimens of these local breeds and in many cases to alter their general characteristics to such an extent as to render it difficult to recognise them.

For instance, within the last twenty years numbers of large
and somewhat coarse pigs of a white colour, with blue spots,
or patches on the skin, coarse curly hair, thick skin, lop ears
and strong and round bone, were to be found in the Fens of
Cambridgeshire, Lincolnshire and Norfolk.  These were very
prolific and hardy, grew to a great size and if given plenty of
good food, over a long period, they would produce a heavy
weight of pork, which was considered suitable for the
labourers, horsemen, shepherds, &c., who were boarded in
the farm houses.  The form and character of these Fen pigs
have changed considerably of late years, owing to the use
of Large and Middle White boars, their quality and fatting
properties have been much improved, whilst they will still
furnish enormous bacon pigs, although their early maturing
qualities are much more noticeable.  The necessity for these
immense and coarse pigs, which were quite common in the
districts named until a comparatively recent date, does not
now appear to exist, or rather the conditions of the life of the
farm hands have so altered that the latter now look for something more toothsome and more easily digested than the very
coarse and heavily salted dried pork, or as it was called by a
stretch of the imagination, cured bacon.  The very low price
at which imported beef and mutton can now be purchased
has also, doubtless, had its effect on the demand for those big
coarse pigs of the Fens.

A far better type of pig, and one more profitable withal,
was the old fashioned plum pudding, or black and white
spotted pig, common in Northamptonshire, Leicestershire and
Oxfordshire.  Some of its many admirers stoutly affirm that
from this mainly, and a cross of the Tamworth and the Neapolitan, and more recently the Small Black, the modern Berkshire was evolved, whilst within the last fifteen years a
serious attempt was made to claim this spotted pig for a
county breed.  Classes for it, even, were established at one, at
least, of the shows in Oxfordshire, but these did not prove a
complete success, as the majority of the winners of prizes

were simply manufactured spotted pigs, being a first cross between the Tamworth and the Berkshire, or a spotted pig and a Large White, the coloured pigs only of the last cross being exhibited. The peculiarly striking likeness of some of the cross Berkshire-Tamworth pigs to the Berkshire pig of some thirty-five to forty years' since, confirmed the Midland Counties' farmers in their view as to the original source of the modern Berkshire, whilst the bacon-curers do not hesitate to assert that the pigs from the Berkshire and Tamworth cross are more suitable for their purpose than the pure-bred pigs of either variety. This opinion is also held by many farmers, who assert that the crosses grow faster when young than the Berkshire pigs and feed quicker, and at less cost, than the Tamworths. In these cross-bred pigs we have the foundation for still another pure-breed, à la Poland China; or do these crosses partake more of a reversionary character?

The breeders of these spotted pigs can fairly claim for them that they are prolific, good sucklers, hardy, good rustlers, and make splendid pigs for cottagers, as they have considerable storage capacity; but their fattening qualities are capable of being improved. In many parts of the Midland Counties these pigs have been crossed with Berkshires until many of them have become nearly all black, whilst others have been so mixed with the Tamworths that the young stock are frequently of a rusty colour; the pigs of this latter tint are thought to be more hardy, but with less aptitude to fatten.

At one period Essex led the van in the breeding of pigs of a small size. A comparatively few Small Whites from Yorkshire or pigs of a similar character and breeding direct from other districts, were kept mainly for exhibition purposes; but these were but little used for crossing purposes with any degree of success, except on the flat sided, strong-haired pigs of Norfolk and Suffolk, on which an immense improvement in the feeding properties was made. The latter county also had its breeders of Small Yorkshires, as they were termed in the sixties. The names of Crisp and Stearn will be handed down to posterity as amongst the most successful of the many exhibitors of the Small White pig. At a period anterior to that mentioned, there existed in Essex and a part of Cambridgeshire, a variety of pigs curiously marked, being, as it was commonly called, sheeted or saddle-backed, the actual colour being a black with a streak of white which extended from behind the shoulders to about the hips. For these pigs it was claimed that they were prolific, good mothers, and quick feeders when put up to fatten. By some persons it is asserted that these sheeted pigs formed one of the foundations of the quick feeding Essex or Small Black pigs, and that another cross used was the black Neapolitan, a pig of another colour, but of a very similar character to the Chinese, which, too, was used to increase the feeding qualities and the prolificacy of the small white as bred in the Eastern Counties. It is extremely doubtful if any material and permanent benefit accrued to either of our English varieties from this cross, as the breeding of our own pigs had been carried on in such a way as to render them mere converters of corn into meat, which consisted mainly of fat or lard. This quality the Chinese and the Neapolitan pigs possessed in an extraordinary degree. One may occasionally see a few specimens of the sheeted pig in Essex, but these are simply the out-croppings of the old breed, as they have ceased to be bred to those points which were at one time considered to be distinctive of the sort.

The common country pig of Norfolk and that part of the county of Suffolk adjoining, has for many years borne a very indifferent character. It is usually a lean, flat-sided, white pig with coarse, harsh hair, and a very slow producer of lean, hard meat. Why it has not been more generally improved by the Norfolk farmers, who claim, or used within a very recent period to claim, to be the first farmers in the British Isles, and of course in the world, is not at once apparent, unless it be

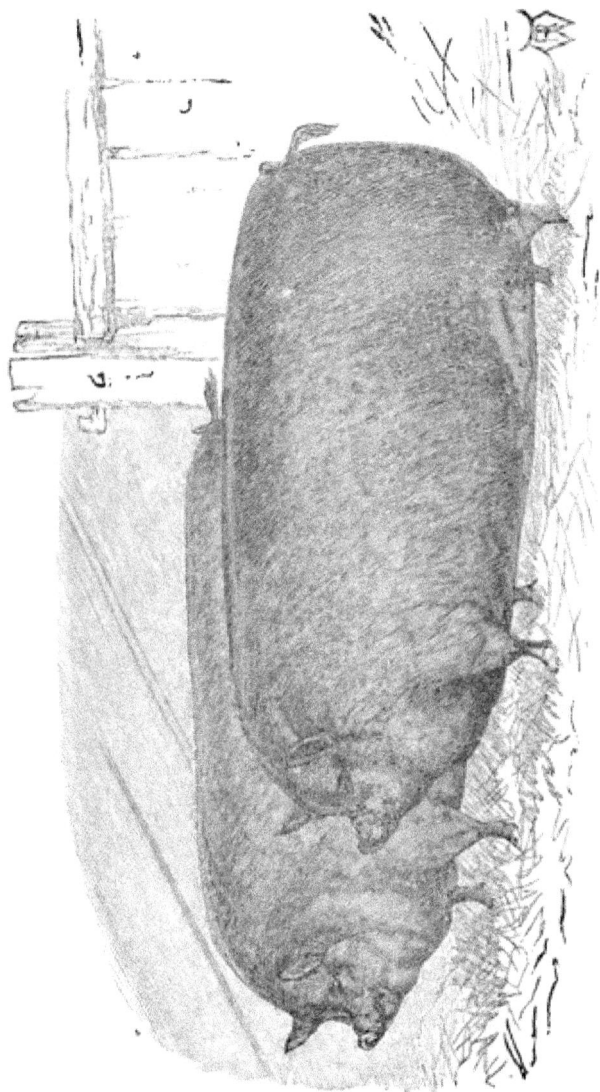

SMALL BLACK PIGS.

Champions at Smithfield Club Show.

nou
po
sy
th
fa
be
to
T
fa
o
c
a
]

owing to the fact that the despised small things, such as pigs and poultry, have been considered of too little moment for profitable production; or it may be due to the fact that the Norfolk farmers, as a rule, have been in the habit of buying their store stock instead of breeding them, and thus have not had their attention directly drawn to the enormous importance of breeding only from the best. The four-course system and the small amount of rainfall are, perhaps, two of the causes of the neglect of the system of breeding and fatting, a system which will more and more force itself into being adopted in the county of Norfolk, not only with relation to the breeding of an improved kind of pig, but also of cattle. There is little doubt that when the time arrives for the farmers in Norfolk to seriously take in hand the production of pigs and profitable pork the change in the pigs in the county will be both speedy and thorough. If we may be allowed to hazard an opinion, it is that the conversion of the Norfolk farmers to the system of combined breeding and fatting of cattle and pigs is nearer at hand than is generally supposed. It is true that the cultivation and improvement of the county pig have been somewhat retarded by the non-success of the bacon factory started in West Norfolk, but this want of success will prove of benefit ultimately, as it has clearly brought out the unacknowledged fact that the Norfolk farmers have something to learn in the breeding and feeding of the correct kind of pig to suit the present market demand.

In Sussex is found a local variety of quite a distinct type and character. It is somewhat difficult to obtain any information of much value as to its supposed origin and the continuance of its existence in its present form. It has a strong local reputation for hardiness, prolificacy, and the furnishing of pork of the kind most in demand in the county. Its colour is locally termed black, but we should call it more of a blue or slate colour; it is of fair length of body, has short legs, rather coarse in bone; it is short of hair, it does not carry itself well, nor can it make good its claim to much style. Its colour, formation, and character generally leads one to believe that the Sussex pig has in its veins a good deal of Neapolitan blood. If this surmise be incorrect, then we are driven to the conclusion that the Sussex and the Neapolitan have many points in common. Some few years since an attempt was made to render these pigs fashionable on the ground, mainly, that the pork from them contained a large quantity of lean meat; but the attempt failed, partially on account, perhaps, of other breeds of pigs—more handsome, or less plain—possessing the same qualities of lean meat producing. The Sussex pigs are said to be very suitable for the production of fat pigs of some 80 lbs. dead weight for the supply of the London, Brighton, and other markets. This we can readily understand, but we venture to remark that in our opinion a cross with the Berkshire or the Middle White would render them more profitable to the feeder without any loss, if not a gain, in the quality of the meat. But here we are met, as everywhere else in the British Isles, with the difficulty of giving any correct data as to the feeding qualities of the various pure or cross breeds of pigs, we mean as to the quantity of food requisite to produce a stone of pork, whilst the entire absence of any properly-conducted block test, or even the knowledge of the mere gross weight of the fat pigs exhibited at most of the shows, including the Smithfield Club's exhibition, makes it extremely difficult for the public and the majority of the exhibitors themselves to acquire any information of benefit from the fat stock shows. In many cases, even the judges appointed are unable to form any estimate of value as to the probable carcase weight of the pigs; and as to the proportion of lean to fat in the carcase, it seldom enters into their calculations. Far too frequently the prizes are awarded to the handsomest and fattest pigs of the various breeds shown, whilst the champion or special prizes most frequently go to unduly fat and wasteful specimens of compact and smart form. There can be little doubt that two of the much-needed improvements in the exhibition of pigs are the weighing of the animals and the publication of the weights and the block test. We shall then have disposed of much of the guesswork which at the present robs our shows of fat pigs of the great proportion of their educational value.

The Hampshire pigs may have had the same source of origin as those in the adjoining county of Sussex, but the breeders of them appear to have been more inclined to produce a pig of a darker colour, somewhat higher on the leg, and of a larger size. It may be that this type of pig was common in both counties and that a different source was sought by the improvers of it for material to carry out their designs. The Hampshire pig reminds one somewhat of the large black pigs found in the West of England, whence perhaps stock pigs have been brought. Again, in the south of the county there is to be found what is locally termed the Improved Hampshire; this is a pig of far more style, a far better coat and more of it, and altogether a more compact pig, indeed just such an animal as an old hand would expect would result from the introduction of the blood of the Berkshire or of a mixture of Small Black and Berkshire. Of course, it is not asserted that this Improved Hampshire has been produced in this manner, and not by selection alone; but it certainly might have been so manufactured much more quickly than by selection alone, even if it must be admitted that the marked change in the form and character of the unimproved pig could possibly result without any introduction of alien blood. Whilst on this subject we may and do frankly admit that a marvellous improvement in the quality and the feeding properties of a given breed may be made by breeding only from those pigs which possess these points in a marked degree, and even the size of the pigs experimented upon may be varied greatly; but to us it appears an impossibility to so completely alter the form and characteristics of a breed of pigs until it resembles in many points the pigs of

another distinct variety without an infusion of the blood of
that variety being introduced. We are aware that it has
of late been publicly claimed that certain distinctive points
common to one breed of pigs have been produced by selections alone in pigs of a totally different and distinct breed in
which these said points have for at least three parts of a
century been wanting or unobservable. Of course it is not
in our province to positively declare this to be an impossibility, but we feel justified in asserting that it is contrary
to our experience, and that the change in form and character
could have been made quite as effectually and infinitely more
expeditiously by crossing and then selecting for breeders only
those pigs which were of the correct colour and yet possessors
of enough of those points sought to be implanted in the
original stock.

Travelling in a westerly direction we find the Dorset,
another slate-coloured and almost hairless pig, with drooping
ears and short legs ; it is credited with being both prolific and
easily fattened. The last quality was considerably increased
by a few careful breeders, some twenty or thirty years since, by
an infusion of alien blood, whether of Neapolitan or of Small
Black, or of both we cannot definitely state, but we are
inclined to think, from the results, a cross of some pig containing many of the qualities of the Neapolitan pig, particularly the aptitude to lay on fat, and the fine sparse hair.
The compound pig was then dubbed the Improved Dorset,
and as such succeeded in winning prizes at the Smithfield
Club's Fat Stock Shows. Mr. John Coate, Ham-moon, has
been a frequent and successful exhibitor of black Dorset pigs
for many years. There does not appear, however, to be many
breeders of the Improved Dorset at the present time, as the
writer vainly endeavoured to execute an order from a gentleman in Uruguay for a boar of some six months old. We have
recently seen a number of very serviceable fat pigs, the result
of a cross of a Berkshire boar with a Dorset sow ; these pigs
were of a darker colour than the sow, and retained much of

the character, but were longer in the body, and appeared to possess a much greater proportion of lean meat.

In the county of Cornwall there used to be found specimens of the smaller and the larger kind of black pig. The former were similar in character to the Small Black pigs, of which some very good and successful herds were kept a few years ago; but the establishment of a bacon factory, and the diminished demand for small and fat pork, have together caused the pig-breeders in the county of Cornwall to purchase a considerable number of Large White pigs, which are found to cross admirably with the Large Black pigs common in most parts of the county. Some of these Cornish pigs are large, having flat sides, heavy shoulders, and lop ears. They are very hardy and prolific, and grow quickly when young, whilst the cross with a really good Large White appears to increase the weight and substance of the hind-quarters and to reduce the weight of the forequarters, and at the same time to add to the proportion of lean meat in the carcase generally.

The pigs usually found in Cheshire and the greater part of Wales are white in colour, and by no means of a profitable character; their harsh coat too clearly indicates their slow fatting, nor can we claim for them an extraordinarily quick growth in extenuation of their inability to make rapid progress when put up to fatten. Efforts are being made by several landowners to improve the quality and the early maturing properties of the county pigs, but at present the improvement is not very marked. One would have thought that the Welsh farmers were most favourably situated for markets, whilst the considerable amount of dairying carried on should render the rearing of pigs a success without any serious outlay.

In marked contrast to the Welsh pig is that commonly found in Lancashire, where so many pigs have been over-refined to such an extent that lard, rather than lean meat, was the chief production of many of those bred in some

3

parts of the county during the "seventies." Very handsome, but almost useless, Small Whites were by no means uncommon, whilst many of the so-called Middle Whites were of a similar character. A decided improvement has taken place in this respect of late. Remarks of a very similar character will apply to the county after which the chief breeds of white pig have been named. Leeds, Bradford, York, and the adjoining towns and district at one time held the field for the production of these white pigs. Some twenty to forty years since, a visitor to any of the hundred and one local shows which appear to be held, would find exhibited large classes of pigs brought out in the best possible prize-winning form. It is true that some difficulty would have been experienced in determining with any certainty to which of the three sub-divisions many of the exhibits really belonged, owing to the far too common habit of indiscriminate crossing, but as individual specimens of the pig which would readily fatten on a minimum of food they certainly stood high on the list. This haphazard system of breeding was greatly encouraged by the multiplicity of small shows, at which prizes were offered for the best pigs of a white breed, so that any pig of that colour could compete, and various means were taken, according to the fancy of the exhibitor, to produce the best of the colour, irrespective of breed or size. Frequently the most successful pigs in pleasing the eye of the judge—who, by-the-bye, was generally a local man who had been the fortunate owner of a few winners of come-by-chance pigs—would be some bred on no particular lines, and often a cross between two pigs of a totally different style and character; thus would often be produced a litter of pigs from which prize-winners in the classes for Large, Middle and Small White breeds could be selected, whilst such an extraordinary occurrence has been known as for one pig to have a prize awarded to it at three of the " Royal " shows, as being the best of its kind and year, when shown in each of the three breeds. For one of these haphazard pigs to commence its successful

exhibition career as a Small White, and to blossom forth as
an aged winner in the classes for Middle Whites, was by no
means an uncommon occurrence, whilst the successful Middle
White frequently developed into a prize-winning Large White,
and, still more strange, in the good old times we saw a sow
win a first prize at a " Royal " show as a Middle White, after
the same set of judges had awarded the first and second
prizes to two of the sow's produce in the class for Large White
boars, and, what is still more remarkable, both sow and sons
were of much the same style and character, or rather of
no particular type; but they were all fat, all white, and all in
the best of show condition.  The educational value of such
exhibitions is not apparent to the ordinary mind.

Cumberland hams were for many years noted, and held a
high position in the market before the taste for small joints
and hams and the mild curing system came into vogue.
These hams were the product of a white pig of more than
the average size and substance; it was generally fine in the
skin, very thin in the hair, neat in the bone, of fairly quick
growth, and readily became fat when it had attained a good
age.  For the present demand, the Cumberland pigs are
considered to carry too large a proportion of fat to the lean
meat; this can, and doubtless will, soon be remedied, as many
of the pig-breeders in that county are thoroughly practical
men, or they would not have so carefully studied to procure
in their pigs the many good qualities which they undoubtedly
possess.  The sows are very prolific and capital mothers, being
good milkers and very gentle.  It is not improbable that this
over-refinement noticeable in some of the Cumberland pigs is
partially the result of too close breeding, at least we have
heard it so remarked by keen observers and by experienced
breeders of pigs who have not entirely confined their attention to one variety.

At one time, within the memory of the author, the county
of Bedford possessed a local variety of white pig, whose origin
was shrouded in doubt.  So far as can be ascertained, it

appears to have had for its parent stock a pig much after
the type found in the Fens in Cambridgeshire, crossed with a
smaller breed of fine quality.   Perhaps the out cross might
have been a Yorkshire of the smaller breed, such as were bred
at the Home Farm at Windsor, at Lord Radnor's at Coleshill,
or in Suffolk by Stearn, Crisp, and two or three others.   The
establishment of a herd of large white pigs by the Messrs.
Howard, quickly had an effect on the ordinary pig of the
county, with the result that some twenty years since the
Bedfordshire pigs were favourably known far and wide, and
led to the extensive trade in store pigs from the town of
Bedford.

To enable our readers to form an opinion for themselves
as to whether their own or their neighbour's pigs most
nearly approach perfection, according to the estimate of our
American cousins, it may be advisable to give here the
results of the labours of a committee of experts, who were
appointed by the American National Swine Breeders' Con-
vention to report on "a general standard of excellence for
a hog which shall best meet the requirements of the market, the standard adopted to represent a perfect hog most
profitable to the farmer and consumer and above or outside of all breeds."   The committee reported that "such a
hog must have a small short head, heavy jowl, and thick
short neck, ear small, thin, and tolerably erect, but it is not
objectionable if it droops slightly forward.   He must be
straight on the bottom from the neck back to flank, let well
down to the knee in the brisket, and possess good length
from head to tail; back broad and slightly curved or arched
from the shoulder to the setting on of the tail; ribs rather
barrel-shaped, tail small.   The hams should be long from
the back to the letting off at the loin and be broad and full;
shoulders not so large, and yet sufficient to give symmetry
to the animal; hair smooth and evenly set in; skin soft
and elastic to the touch; legs short and small, set under
the body, and the space between wide; a good depth

between the bottom and top of carcase. He must be possessed of a good, quiet disposition, and as a general rule should not weigh more than 300 or 400 lbs. gross at twelve to eighteen months. Colour may be black or white, or of a mixture of the two. Such a hog will measure as many feet from the top of the head to the setting on of the tail as he does round the body, and as many inches around the leg below the knee, as he does in feet in length, or around the body, and the depth of the body will be four-fifths of the height."

It is not suggested that a pig such as described would fill the bill on this side of the Atlantic, nor must it for one moment be assumed even that we look upon the described perfect hog as exactly the kind of animal most profitable to the consumer; and if it be not to the consumer and to the middleman, it will not long continue to be the most profitable to the producer. There are several points mentioned as necessary which, to say the least, may be looked upon by practical men as partaking very much of those characteristics which in this country would be called fancy points, whilst others, such as the heavy jowl, short thick neck, &c., would be considered most objectionable to the consumer and unprofitable to the feeder of the pig.

# CHAPTER II.

## SELECTION OF THE BOAR.

The choice of the sire is one of the most important of all the steps to be taken in connection with the successful breeding and rearing of any of our domesticated animals. It may truly be asserted that with pigs a false step can be more quickly retraced than with animals such as the horse, or even with cattle, since the error can be more readily discovered, owing to the shorter time taken in their reproduction; but this, to our mind, does not in the least lessen the urgent necessity for the greatest possible care being taken when starting a herd of pigs, or when purchasing a boar for use in one's own herd and for the benefit of those of one's poorer neighbours who breed pigs and often trust to the returns from their one or two sows, or from the fatting pigs, to pay their cottage or allotment rents.

The loss which often arises in a district from the introduction of an inferior boar is very great, and frequently affects most injuriously those whom it is the endeavour of the owner of the boar to benefit. The mere fact that the produce of a boar may, and often does, amount to hundreds of pigs in the course of only a few months renders it desirable that care should be taken to select one preferably from a breeder of repute and whose stock have stood the test and have been favourably known for a lengthened period. It may be remarked that a herd of so-called pedigree pigs can be got together from all parts of the country with comparative celerity,

BERKSHIRE BOARS.

The property of Sir Humphrey F. de Trafford, Bart.

whilst the number of local and county shows at which fattened
pigs of nondescript or mixed breeding can be successfully
exhibited is greatly owing to the small value of the prize
money not making it worth the while of owners of really good
pigs to incur the heavy expenditure for entry fees, subscrip-
tions, cartage and railway charges, to exhibit their pure-bred
swine.    This reluctance is also enormously increased by
the ofttimes ridiculous decisions given by the judges who,
having a good knowledge of some other variety of stock, are
selected to make the awards in the classes where the stock
which they may understand are exhibited, and then have the
judging of the pigs thrust upon them, so that the cost of
having a practical judge of the pigs appointed may be saved
to the funds of the society.   We have in our time seen some
most extraordinary and ridiculous decisions given in connection with the awarding of the prizes offered for pigs, decisions
which would be ludicrous, indeed, if it were not for the
injustice meted out to the unfortunate exhibitors by those
judges who lightly undertake a duty of which they are often
totally ignorant ; and, as is frequently the case where there
is ignorance, prejudice also is strongly *en évidence*.   The mere
fact that a person has the temerity to undertake public duties
beyond his capacity is not often a proof of the acceptor's
good sense, but rather that he has egregiously failed to place
a just estimate on his abilities.   Another source of annoyance
and hindrance to the exhibitor of pigs, and a great difficulty besetting one who is desirous of improving the swine in
his district by securing a good exhibit of a variety new to the
neighbourhood, is the local judge who will persist in believing,
and in acting most determinedly on his belief, that "whatever
is, is right" in connection with the style of pig in his own district.   An amusing instance of this occurred at a county show,
far from the happy hunting grounds of the White pig.   In
that county a most praiseworthy effort, and one bound in
the end to prove a success, was made to furnish the
local farmers with a good market for the disposal of the

fat pigs which they were compelled to produce in order to
successfully and profitably carry on the dairying industry
which was most general in the neighbourhood. A company
was formed to equip a bacon factory with all the most modern
and expensive accessories, and steps were taken to give practical proofs to the local farmers of the desirability, and even
the urgent necessity, in the interests of all concerned, that pigs
of a certain age, form, and state of fatness should be sent to
the factory for conversion into bacon, &c., and that it was impossible to carry on the concern to their mutual benefit unless
the best class of cured meats were manufactured, since the
inferior grades could not in England be made to successfully
compete with bacon of this class produced so extensively
abroad. The best obtainable and most practical pamphlets
and diagrams were distributed broadcast in the district, and
a few of the correct type of pigs were introduced ; but the
major part of the pigs sent to the factory continued to be of
the undesirable type—too fat, old, and much too heavy in the
forequarters, and not long enough or deep enough in the
sides. It was then decided to make a further effort, and to
offer at the county show a number of prizes for the variety of
pig which has been proved all over the world to be the best for
the bacon-curer's purpose, either when bred pure or crossed
on the local breeds. Although the show was to be held some
four hundred miles away, the writer acceded to the wishes of
the projectors of the scheme to send a number of his Large
White pigs to compete for the special prizes, as well as in the
classes where the prizes were offered for the best pigs under
twelve months old, irrespective of breed or colour. He entered
a young Large White boar, which had previously won in
Yorkshire, and a young sow which had won a first prize in
a strong class at the Otley show. Other White pigs were
entered, and an entry in each class was made by a local
exhibitor who, to compete in another class, sent a useful black
sow, heavy in forequarters and with lovely lop ears which
completely shut out the daylight. This sow had on her a

litter of pigs of an exactly similar style to herself, and, having only sows of the same character to compete with, she won the first prize, being aided by her large litter. Then came the amusing part, as one of the boar pigs and one of the sow pigs of the litter were moved into pens to compete in the classes for pigs under twelve months old and against our ten months old pigs and previous winners. The two judges —both local men—looked very wise, and these two little suckers were respectively placed first in the two classes, the two pigs shown by the writer being placed second to them. A gentleman who had had great experience with pigs asked us what it meant, to which we replied, "Ask the judges." He did so, when they frankly admitted that the pigs placed second were the best they had ever seen, but that the people in the district liked the black pigs better than white ones, so they awarded the first prizes to the black ones. We could not repress a smile nor the remark that "we admired the honesty of the judges, but had grave doubts as to their wisdom." Thus are the best efforts to improve the common stock of the country by importing good sires sometimes partially frustrated by local prejudices! This peculiar and —we should hope in more senses than one—singular case gives some idea of the difficulties that would-be benefactors meet with, and yet the laudable attempts continue.

The oft-quoted remark that "the bull is half the herd" has at least equal appropriateness if applied to the boar and the herd of swine; it is therefore most essential that as many good points as possible should be embodied in the stock boar. The first essential undoubtedly is that he should be of a pure breed—that is, one which has been bred for a certain number of generations on certain lines or a standard more or less defined—and in order to be certain of this, and of the fixity of those good points of the boar which are apparent, an old-established herd should, as a rule, be visited and the choice of the boar be made after seeing the parents or other relations of the boar. Our reason for advising great care in making the selection, is that in a newly-established herd, even though the owner may have won prizes with his pigs at the recent shows, a great variety of type is likely to be found to exist, and more frequently than not the prize animals may have not been bred by the owner of the herd, but bought from several other herds of greatly varying type. These animals may themselves be of prize-winning form and character, yet their produce, when mated together, will often be very unlike their parents or each other. The mere fact of animals being owned or occasionally bred by an exhibitor does not necessarily stamp his herd as one of general excellence, although it has in times past been the fashion with many amateurs to immediately flock to make purchases from a herd which has been successful at a few shows, with animals mayhap picked up at a great price all over the country. Fortunately for the best interests of stock-breeders, the folly of this proceeding has been made patent.

It is a generally-accepted opinion that the male animal exerts a far greater influence on the external points of the joint produce than does the female parent, the latter in turn influencing the internal portions to a greater degree. It therefore becomes necessary that in selecting a boar one should be sought for compact in frame, as long and deep in carcase as is possible, consistent with strength; well developed in the hind quarters, with a wide chest, ribs well sprung, deep flank, legs placed well outside the body, bone fine, pasterns short, neck muscular and fairly long, head of medium size, but wide between the ears and eyes, the latter being bright and lively, indicative of sexual energy, and all covered with as much fine, straight, and silky hair as possible.

It is scarcely to be expected that very many boars of any pure breed should be found possessing the whole of the points enumerated, but search should be made for one as nearly as possible answering to the model boar as described. It may be necessary to call attention to the necessity for the sexual organs to be well developed, as this is frequently an indication of constitutional vigour. Again, never be tempted to use a boar suffering from rupture, or one having but one testicle visible, or as these boars are generally termed, *rigs;* both of these are serious defects and certain to be transmitted from parent to produce. Of the two, perhaps the rig pig is the greater nuisance, as it is almost impossible to thoroughly operate on it, and the result is that the meat from such a pig is as strong and high-scented as is that from a boar pig, and besides this a partially-castrated pig is generally a restless and unthrifty one. Until within the last few years the ruptured pigs were not usually operated upon, but converted into sucking or porket pigs, and thus killed before the flesh had become tainted; but now the castrator generally operates on these pigs, and afterwards, by means of a few stitches, sews up the incisions made in the scrotum or purse.

A vicious and bad-tempered boar should be shunned as though it were a plague-stricken animal; it is a most dangerous brute, as it will attack its feeder or any animal which it may encounter during its fits of anger, and these are frequent and arise suddenly, and without any apparent reason. Even when the tusks of a vicious boar are broken off it is capable of inflicting severe wounds, which are often most difficult to heal. Many a good animal, particularly mares with foals and cows with running calves, have been disembowelled or seriously injured by vicious boars having been allowed their liberty, or having broken bounds. Besides this insatiable desire to injure human beings or other animals, a bad-tempered boar is certain to transmit this irritability of disposition to his produce, which will prove restless, unsettled, and very slow feeders. A great point in pig-fatting is to secure pigs which are naturally inclined to rest in a state of contentment from one time of feeding to the next; these pigs will extract a far greater proportion of the nourishment contained in the food than will those restless pigs which bark and gallop round the sty at the slightest unusual noise.

Some years since the writer was very much surprised to

notice at shows that many of the pigs exhibited were possessed of deficient udders, in that a number of the teats were what by old pigmen is termed "blind"—that is, the nipple is not prominent—so that when the little pig essays to grasp it with its lips, the teat recedes and the pigling meets with disappointment instead of a supply of the nectar of life. We also noticed that a number of the boars showed signs of this malformation of the teats; this led us to make enquiry as to whether or not this failing was hereditary, and the general opinion appeared to be that blind teats, like most other malformations, were in most cases the result of breeding from boars or sows or both of the parents having defective teats. We quite believe in this, and should not hesitate to cast the best-looking and most stylish boar had he any of these defects in the shape of blind teats.

It may to some persons seem strange to make this so strong a point when selecting a boar, but we have had so many proofs of the damage done to a herd through the neglect of such apparently trivial matters—if we may class the formation and usefulness of the udder in this category—that we deem it desirable to lay particular stress on the number, position, and apparent utility of the teats of the boar. In this, as in most other parts of the form of its produce, will the impressiveness of the boar be noticeable.

It is equally advisable to select a boar which is one of a large litter, as prolificacy is just as hereditary through the sire as through the dam, as by so doing the chances are in favour of the boar begetting large litters of strong, healthy pigs. Although some persons make mere size a great point when choosing a boar, our experience leads us to consider this to be a mistake; a very large boar seldom lasts long; he becomes too heavy for the sows; he probably proves to be slow, and his litters few and small in number. A very large and heavy boar is also more likely to suffer from weakness of the spine or hind-quarters, and is frequently weak in his joints and bent-legged. These latter failings should be

LARGE WHITE BOAR, BEACONSFIELD.
Winner of First Prizes.   Bred and exhibited by Mr. John Barron.

specially avoided, as they are most hereditary, and will frequently crop up for several generations. Weakness of ankles and roundness of bone—two qualities which should be avoided in a sire—are often allied with great size. A medium-sized, compact boar, heavy in the hind-quarters, and light in the fore-quarters will frequently continue fruitful for at least twice as long as will the heavy-shouldered and coarse-boned boar. Nearly the whole of the most successful stock pigs have been on a small rather than on a large scale, whilst for exhibition purposes an overgrown, ungainly animal is certain to be a source of disappointment, as, if it be possible to train one of this character, a very small amount of railway travelling, excitement, and worry will render it unfit for showing, as the training and feeding will most probably have rendered it useless for breeding purposes.

One often hears the remark that boars should be wide in the shoulders and have thick necks, whereas these are two points of very little importance, even if it be not advisable to avoid them, as they are not indications of high quality of meat, nor, with the present fashion or style of carcase required by the butcher, the bacon-curer, and the public, are they approved of by those who eventually place a value on the meat of the fat pig.

# CHAPTER III.

## MANAGEMENT OF THE BOAR.

THERE is little doubt that the boar is one of the most important members of the herd, even if it be not, as it is very frequently asserted to be, half the herd. It is therefore desirable that especial care should be taken with its management, as a grievous loss to the owner of a number of sows may result from want of proper care and attention being paid to it. How great the loss is known only to those who, having secured, after much trouble, a really good boar—one docile, prolific, and the sire of good pigs—are unfortunate enough to lose its services. Our own experience leads us to the belief that nothing in pig-breeding gives more cause for anxiety, and necessitates more caution and study of points and pedigree, than does the selection of a boar to head one's herd. It therefore follows that the loss of such an animal is something approaching a calamity to one who is building up a herd, and is particularly annoying if by any means the loss of the services of a good boar has arisen from neglect of any kind. There are various causes, besides the actual death of the boar, which may have the same result, but to these we will refer as we proceed.

We hold strongly to the belief that it is advisable to select the boar when it is young; if it be possible to see it on its dam, and to also see its sire, so much the better. We will therefore assume that the delivery of the boar follows the weaning of it. It should be placed with other pigs of about the same age, and fed generously on sharps, with a little meal

added; this two or three times per day; and then between the
morning and the midday meal give them a few peas or a
little whole wheat or oats. If skim milk can be obtained,
the young pigs will pay as much for it as will any other kind
of stock, and further, they will make far greater progress
with than without it. A run in a small paddock or enclosure
for a few minutes each day will help to keep the pigs in
health. This system may be followed until the young boar
is some five months old, when any unspayed sow pigs in the
lot should be weeded out; the food may be increased in
richness by the addition of more meal, without any fear of
the boar becoming too fat, provided sufficient exercise be
allowed; this last is most important, as good feet and ankles
can only be retained in this way, and further, as soon as the
boar has been used, it is seldom possible or advisable to
allow him much liberty.

Opinions vary much as to the age at which a young boar
should be allowed to commence service; practical men may
be found who would strongly recommend that it should be
some nine or ten months old before being used, whilst others
of equal experience advise that seven or eight months is the
better age at which the boar should commence work. There
may be much to be said in favour of both these views. The
breed of the boar is thought to affect the question; for
instance, a boar of any of the smaller breeds, such as the
Small White, the Small Black, or the Berkshire breeds, might
be thought to be fit for service at an earlier age than one of
the Large White or Middle breeds, as the latter would require
more time to develop. In practice this is not found to be
the case, as the Large White boars are quite as precocious, if
not more so, than those of the smaller varieties. Again, the
system of feeding the young pigs would have a great influence,
as, should they be kept in high or good breeding condition all
their young lives, the boars would be more able to withstand
the strain of the use of their procreative powers when eight
months old than would other pigs of a similar character

which had been very sparingly fed until they were ten months old.

Some persons have an idea that the breeding from comparatively young pigs has a tendency to shorten the period of their usefulness. This is certainly not in accord with our own experience, as some of our boars which were used before they were eight months old have continued fruitful until they reached the age of seven or eight years. As an example, one of the best boars, both individually and as a getter of the very best pigs, Holywell Jimmy by name, a boar selected some years since as the best representative of the breed then extant, was sold by the writer when about six months old. The boar went into hard work at once, was used alone in a large herd of sows, and after he was brought again to his birthplace large numbers of sows were put to him, and he continued vigorous until at the age of over eight years when he met with an accident which resulted in his being slaughtered. Again, the Large White boar Holywell Squire II. was farrowed in the month of January, 1888, used before he was eight months old, and continued in work for some months, when he was prepared for the 1889 summer shows, when he won prizes at the shows of R.A.S.E. at Windsor, the Highland Society, the Bath and West of England, &c., and was shown successfully for two years afterwards, winning, amongst many other prizes, the first prize two years in succession at the Yorkshire Agricultural Society's show. After his most successful show career, he was used very extensively, and begat hundreds of pigs, several of which proved successful in the various showyards at home and abroad. When still quite fruitful, at six years old, he was injured by a sow, or at least when one was put into his pen for service; this necessitated his slaughter.

Two other remarkable instances are now being used in our herd at the time of writing; these are Holywell Dublin and Holywell Count. The former was farrowed on September 15, 1889, and was shown with his dam and his brothers and

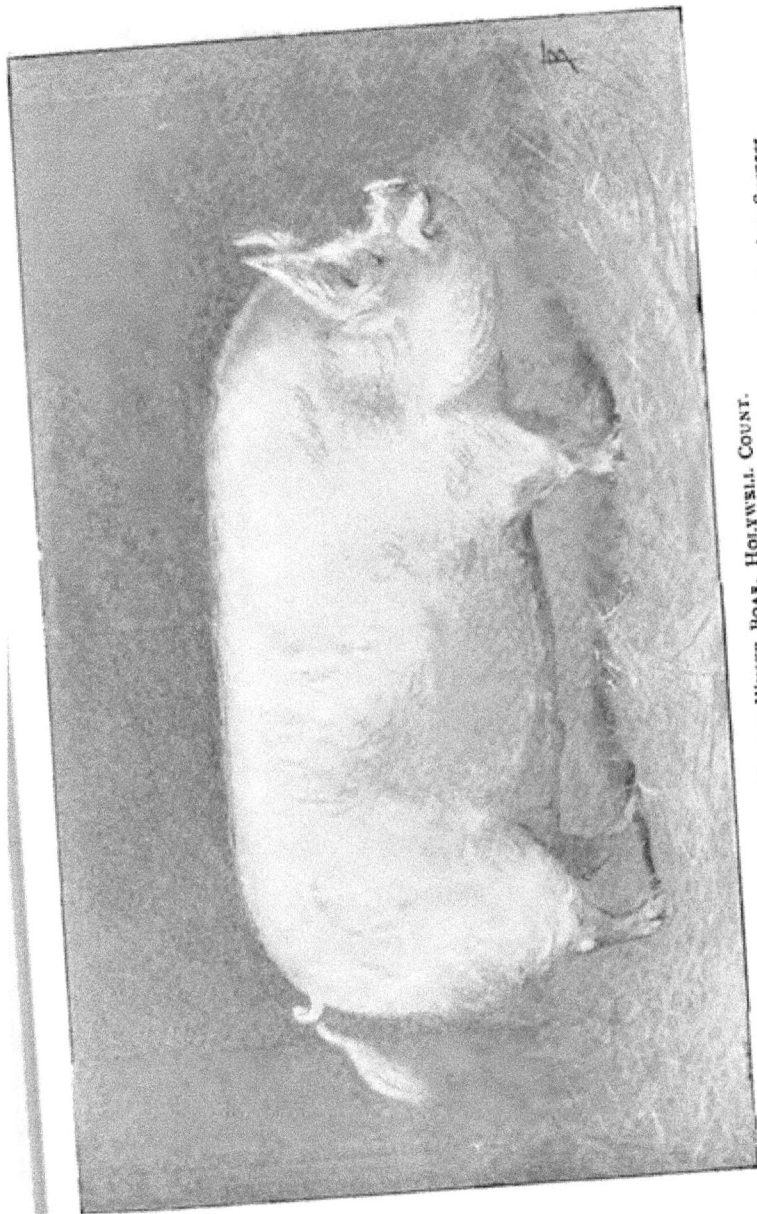

MIDDLE WHITE BOAR, HOLYWELL COUNT.    Bred and owned by Mr. Sanders Spencer.

Winner of Twenty-five Champion and other Prizes.

sisters at the Tredegar Show in the November following, when the first prize was won; he was used during the spring of the year 1890, and also shown twice during the summer, winning first prize each time. He was used during the remainder of the year, and in the spring following was sent to the Royal Dublin Show, where he won first prize in the class for aged Large White boars, and was declared by one of the judges to be the best pig he had ever seen; the boar also appeared at the Bath and West, the Highland, Yorkshire, and many other Shows during his career, during which he scored some twenty-five wins; he is now, when over seven years old, working as well as ever. The second boar, Holywell Count, was farrowed in June, 1880, was used the following spring, and when eleven months old won the first prize at the Oxfordshire Show, followed by firsts at the Bath and West, the "Royal," &c. The boar was afterwards used, and has added to his credit some twenty-five prizes and cups, one of the latter being the challenge cup, won last season (1895) at the Lincolnshire Show, when he beat winners at the other chief shows of the season. Holywell Count is now again at work as well as ever, and promises, if he escapes accidents, to be fruitful for years to come. We could cite numerous other instances where early work, even combined with years of showing, has not impaired the usefulness of boars, but we think that the above instances prove our contention.

It is a great mistake to allow the boar to run about with the sows; accidents to the other stock and to the boar are frequent when this is permitted; the boar frequently becomes savage, both with the sows and with the attendants; the sows are filed and the litters of pigs smaller, and not so strong, and the vigour of the boar is impaired from the frequent and harmful service of the sows, which will continue with some sows for several days. The better plan is to keep the boar in tolerably close quarters, and then when the sow has been in use a day or two, have her put into the boar's sty, and

4

removed after one complete service; this will be found to be equally as effective as a dozen matings, whilst the strain on the boar will be infinitesimal, compared to the heating and excitement of being in the company of the sow during the whole period of her œstrum; besides, the boar's services can be made available for many times the number of sows. As soon as the sow is served, the attendant should enter the sty and quietly turn out the sow, being particularly careful, as at all other times, not to use the stick, which he should always have in his hand at such times, unless really necessary. If the boar is kindly and firmly treated, no trouble will arise, providing, of course, that it is possessed of a good temper, and one with a bad temper should never be kept.

As to the number of sows which a boar is capable of serving, or should be allowed to serve, much will depend on the age of the boar, and also on its constitutional vigour; in the latter quality there is very great variation in the boars, even from the different herds. Where soundness of constitution and prolificacy receive the proper share of attention, rather than so-called quality, aptitude to fatten and show points, the boars will not only serve a much larger number of sows, but will be far more certain, and continue to be fruitful a much longer time. A customer of ours in Denmark, who would buy twenty or thirty young boars at a time, declared that these boars frequently had to serve from 120 to 150 sows in a year; this is, in our opinion, a too heavy tax on a boar, which will easily and effectually serve fifty sows per annum, if it be fed carefully, and managed on the system traced out above.

The food of the boar will require to be regulated according to the amount of work which is being required of it, and to its age. A strong, vigorous boar, from one to four years old, may be fed chiefly on swill, stirred with a very little barley, pea, or bean meal, twice a day; the remainder of his food may consist of grass, green clover, tares, or lucerne, or a limited quantity of cabbage in the summer time, and turnips, swedes, kohl rabi, and mangels during the eight months from the end of August to the end of April. A boar will, when not being much used, almost entirely live and thrive on green food. The mangels are best used in the spring, whilst a limited quantity of cabbages, and even at times of kohl rabi, should be given, as both are liable to cause constipation; the former is especially apt to have that effect on pigs which are not allowed exercise. As the age of the boar increases, or when the amount of service required of him is considerable, the quantity of meal should be increased, just so that the condition and vigour are preserved. It is not the fat boar which is always prepared for service and the most prolific, but one in fair flesh and one not used too frequently. We have often noticed that the litters of pigs will be fewer in number and less vigorous when farrowed from a service by a boar which had had two or three sows within the previous twenty-four hours, and, on the other hand, a long rest does not tend to make the boar more useful or prolific. The cause for this is not far to seek: the generative organs are in the most healthy condition when they are in general use.

One sometimes hears the question asked as to the best manner of disposing of an old boar—whether to sell it in store condition, to fatten it as it is, or to have it operated upon before being put up to fatten. An opinion appears to be general that it answers to have the boar castrated, and then to commence fatting it, as it is asserted that all the meat gathered on to the bones will be fresh, and therefore more tender and free from the rank scent of the flesh of a boar. Our advice would be to sell in store condition, and for the following amongst other reasons. The operation of castrating an old boar is a very painful one; it is also attended with danger to the operator and to the boar. The fatted boar will realise but a very low price per stone on the market—frequently not more than 1d. per lb.; the castrated boar, if not injured in the operation, will, when fattened, realise but a very little more per lb., and further, it does not pay to use good meal to manufacture meat which will only realise 1d. or 1½d. per lb., as this is about the value of boar or stag pork. As to the suggestion that such meat should be converted into bacon, no one would eat it when made.

# CHAPTER IV.

## SELECTION OF THE SOW.

If we apportion to the boar the chief influence in the formation of the young pigs, an equally, if not more, important duty falls to the share of the sow, to furnish the body with the necessary internal arrangements to enable the complete animal to readily convert its food, so that the pig grows rapidly, fattens quickly, and proves itself a profitable hog.

Some old breeders exhibit a strong predilection for what they call a big roomy sow; with this view we are in accord, providing looseness and ungainliness of form do not accompany this size, as a more helpless and unsatisfactory brood sow does not exist than one of those unwieldy, flat-sided, weak-loined animals, which some few persons delight to honour at the shows because of their immensity, or it may be on account of the vast amount of wonderment they generally create in the simple minds of those whose knowledge of the points of a good brood sow is very limited. It will not be necessary to the tyro in pig-breeding to witness more than once the frantic efforts of one of these flat-sided giantesses to regain its feet after giving suck to its young; the chances are that the latter will be scattered all over the sty by the hind legs in the struggles of their dam, particularly if the floor be at all slippery, or the back of the sow be at a somewhat lower level than her feet. An evenly-made, compact sow, with quarters long, wide, and deep, and on short legs, will rear far more pigs, and at much less cost, than will

one of the very largest kind.  As to the particular points to
be sought for in a breeding sow these are very similar to
those required in the boar, particularly the gentleness of
disposition, as a barking, savage sow is seldom a good milker,
and besides this there are frequently occasions when it is
absolutely necessary for the pigman to enter the sty in order
to assist the sow in difficult cases of parturition, to remove
the after-birth, and often to take the little pigs away from
her to break off their teeth.

A well-formed udder is most necessary, and no sow should
be bred from unless it has at least twelve teats, and if it has
fifteen so much the better; these should be even in size,
placed equi-distant apart, and commence as near the forelegs
as possible, and the whole of them should be milk-giving
teats.   The blind teats described in a previous chapter should
be avoided, as should those small teats which are generally
placed in close proximity to a full-sized teat, and give only
a little milk; the sucker to whose share one of these under-
sized teats falls is sure to be a wreckling, pitman, or whatever the local term may be for those unfortunate little pigs
which have failed to make as quick growth as their brothers
and sisters.   An opinion appears to exist that the number
of teats possessed by a sow is a sure indication of its prolificacy; our experience leads us to hold a different opinion,
as in our own herd we have had examples of sows with only
ten or eleven teats producing seventeen and eighteen pigs at
a litter.   It appears that the chief reason for a sow which
possesses a large number of teats also producing numerous
progeny is that this particular quality of prolificacy has been
studied at the same time that the formation of the udder has
received attention from the person or persons who have bred
the sow's ancestors for several generations.

We strongly hold to the belief that the good, but somewhat
unusual, quality amongst pure-bred pigs of farrowing large
trips of pigs becomes a more fixed quality of sows of a
particular tribe or family than does that formation of the

MIDDLE WHITE SOW, HOLYWELL BEAUTY II AND LITTER.
Winner of First Prizes.    Bred and exhibited by Mr. Sanders Spencer.

udder which is considered desirable. We can recall instances of sows having but ten or eleven teats and yet for generations their forbears had been possessed of well-nigh perfect udders ; but it is not within our recollection that a single one of these sows did not possess in a marked degree the prolificacy for which their ancestors had been noted. Not only so, but the majority, if not well-nigh the whole, of their produce had well-developed udders, although in the next generation an occasional yelt would be noticed which had but a limited number of teats.

It appears to be quite possible to so develop the powers of reproduction amongst sows that they become too prolific. We have several times experienced this, as young sows of certain families have farrowed from seventeen to nineteen pigs for a first litter. This is far too many for a young sow, if she be but about a year old, as to bring up six or eight bonhams at the first attempt is quite enough of a tax on the really immature dam, particularly if she farrow between the months of August and April. A yelt will rear a greater number of pigs, with far less strain on her system, during the summer than the autumn, winter, or spring months.

We have noticed that the best pigs of a litter are almost invariably those which suck from the teats nearest to the forearm and that the best milking sows are those whose teats extend quite along the belly, some of them even showing between the hind legs.

This quality of milk-giving is not sufficiently studied by many pig-breeders, who take it for granted that well-nigh every sow which will produce pigs will, as a matter of course, furnish the latter with a good supply of lacteal food. This is an entire mistake, as there is nearly as much difference in the milking qualities of sows as in those of cows. Nor does the coincidence end here, since the good suckling sow is in many respects similarly formed as the heavy milking cow—thus both will be light in the neck, comparatively narrow at the top of the shoulders, long in the quarters, and, when in milk,

light in the second thighs; both, too, will have the same
gentle, quiet disposition, indicative of a desire to make the
best of the good things which fall to their share from the
feeding-bin.

A really good milch sow will require rather more care when
her pigs are three or four weeks old and need a more nourishing diet, but to this we will refer in the chapter dealing with
that subject.

We have often been amused with the way in which many
otherwise good judges of stock proceed to select the best sow
pig of a litter for the purpose of growing or for breeding purposes. The very first point for which they seek will be the
thick short neck and wide-topped and often upright and coarse
shoulders, for the reason, as stated by one celebrated sheep-
breeder, who is often called upon to act as a judge of cattle,
sheep, and pigs, " You want weight in a pig, and you must
have wide shoulders in order to get it." A greater fallacy
has never been exploded, nor one which has worked so much
harm amongst the breeders of pigs. A sow such as described
is invariably a poor suckler and generally a slow breeder ;
two of the worst qualities for which the so-called pedigree
pig of some fifteen or twenty years ago acquired so unenviable
and so damaging a notoriety. This very natural objection,
entertained for years to ˙sows of this formation by really
practical pig-breeders, has within the last ten years received
a marvellous confirmation on the part of both bacon-curers
and consumers of pork. The former will not, if possible, convert such pigs into bacon, and the best customers amongst
the consuming public will not buy the bacon when manufactured from fat pigs of this description.

We have thus a consensus of opinion against the heavy-
shouldered, thick, and short-necked pigs ; the intelligent
breeder fights shy of them, the bacon-curer objects to them,
and the most important personage of all, the one who finds
the most money, will have none of them ; and yet, sad to
relate, many of our so-called judges refuse to be educated, and

still delight to honour them. Time and public opinion alone will eventually either convert them from the error of their ways or cause their places to be filled by those who attach a higher value to practical than to those fancy points which have in times past proved a perfect curse to many of the breeds of our domesticated animals. The crusade which set in some fifteen years ago against the growing folly of judging pigs from a fancy standpoint has already borne good fruit ; those helpless animated bladders of lard, which were entirely innocent of all the paternal and maternal duties of pig-life, but which secured the chief honours at the shows where prizes were supposed to be given for breeding-stock, are now seldom seen, although even at some of the shows held during a recent season a few of these useless specimens were decorated with the winning rosettes. Here, again, a pecu-liarity is noticeable ; these over-fatted, bloaty, and thick-shouldered pigs of the larger size were undoubtedly cross-breds, and it is therefore to be devoutly wished simply sports ; and yet these sports and cross-breds being allowed to compete with pure-breds is an injustice to those who persistently confine themselves to pure-bred sires and dams, and more particularly is the unfairness of the system apparent when novices are, as the Americans have it, selected "to tie up the ribbons." The broad fat back, the quick-silver, *alias* lard-producing back, and the carefully washed and oiled hair, act as a charm to the breeders of sheep, who value in the pig the points that are most difficult to attain in the sheep, and in the carrying out of their theory tend to do irreparable injury to the breeding of the most valuable and the most useful style of pig.

Still another quality, that of prolificacy, should be a requisite in the young brood sow, and this can only be secured by selecting the young stock from a breeder who has for some considerable period made this a particular point in his system of pig management and breeding. It is quite possible to secure from a herd bred solely with a view to winning prizes

a young sow which may prove to be prolific and a good
suckler, but the chances are very much against such good
fortune; and it is even within the regions of possibility to
secure a young sow which will farrow large litters and yet
fail to rear a decent litter of pigs, since the quality of milk-
giving does not by any means necessarily accompany prolific-
ness, although the reverse generally holds good. One of the
chief causes of this want of prolificacy in some of the successful prize-winning pigs is the offering of prizes for sows called
breeding sows with or without the qualification that they
must be said to have produced or shall produce within certain
defined periods litters of live pigs. It is quite possible to
make an in-pig sow sufficiently fat to be able to compete
successfully at the summer shows, but it will pass the wit
of man to get up a sow sufficiently to stand any chance of
success against empty or rested sows if she has during the
spring suckled and brought up a good litter of pigs. The
feeding alone would render this impossible, as the food most
suitable for a sow suckling a large litter of pigs is of an exactly
opposite nature to that on which the sow would make fat and
get into show condition. Besides this, not one sow in fifty
which has been forced for exhibition will properly rear any
great number of the pigs which she may be carrying during
her exhibition tour. One of the best remedies for this unfortunate and unsatisfactory state of affairs, would be to offer
the prizes for sows in milk, the condition of the pigs to be
taken into consideration. When shows were first established
there undoubtedly existed a sufficient reason for the attempt
to encourage breeders to rear pigs which had an aptitude
to fatten ; but this has entirely changed, since now the difficulty is to find a sufficiency of highly-bred pigs to grow lean
meat and to rear large and healthy litters of young pigs, and,
what is more, this inability on the part of exhibition pigs to
perform the two most important duties of the commercial
pig has been a source of considerable loss to the country
generally, and to the pig-feeding industry, which has a

good deal of influence on the successful management of the dairy farm, and of the allotments and the gardens of the dwellers in the country.

# CHAPTER V.

## MANAGEMENT OF THE SOW.

THE young sow – or, as it is variously called, according to the district in which its owner resides, elt, helt, hilt, gelt, gilt, yelt, yilt, &c.—should be ready for being mated by the time it is eight months old. Numbers of young sows will show signs of œstrum when they are four or five months old, if highly fed on forcing diet, as many of those pigs which are prepared for exhibition are fed. But for many reasons it is not advisable to allow them to be mated at so early an age, as only a very few yelts of nine months old would be able to satisfactorily rear a litter of pigs. A varying proportion of the youngsters would be found to be stunted and weakly, whilst the whole of them would be under an average in size and substance. Besides this, losses of both yelts and pigs would frequently occur during parturition. Only recently two instances were brought to our notice where the owner carelessly allowed a young boar to run about with a number of unspayed sow pigs; two of these proved to be in pig when the owner, anxious to keep up the strength of the young yelts, and to assist them to nourish their future litters, gave them a little extra food. When the yelts were due to farrow, they were in fresh condition, though by no means fat. The young pigs also proved to be well-grown; but when the yelts attempted to farrow, it was found that the passage was far too small for the yelt to get rid of the pigs without aid; assistance was rendered, and eventually both yelts relieved

A. M. Gauci.

LARGE WHITE PIGS. Winners of First Prize and Breed Cup at Smithfield Club Show. Bred and exhibited by Mr. Sanders Spencer.

of a portion only of the pigs. One of the yelts succumbed during the operation, and the other became so weakened that the little pigs were not brought forward, and the yelt, after being dosed and nursed for a few hours, also died of inflammation and exhaustion. The extra feeding might have caused the little pigs to be larger than they would have been had the yelts been more sparsely fed. But in any case the risk of farrowing down such young and immature yelts is always very great. Besides this, nothing is gained, as, should the yelt bring up a litter without any serious accident, she will have become so reduced in condition that a rest of some weeks would be necessary for her to regain her strength, even if she were not permanently stunted and injured. The pigs, too, would be of less value than would those from a fairly well matured yelt.

It may not be out of place to now mention the fact that an opinion prevails that the pigs of a first litter are of less value for breeding purposes than those from a matured sow. If there were any good grounds for this opinion, would it not tend to prove that those who object to the law of primogeniture had some good grounds for their objections? If the first-born of stock were deficient in stamina, substance, and other qualities equally necessary in man and beast, why should not the same law, if it existed, rule with human beings? So far as our experience has extended, we have not yet proved that the produce of yelts, if farrowed when the dams are fairly well matured and capable of rearing their young in a healthy, thrifty manner, are not equally valuable for breeding purposes as the pigs from sows; some of the very best pigs for both the breeding and the show pen have been bred in our herd from yelts not more than a year old when their youngsters were farrowed, care only being needed that too large a litter be not left for the yelt to rear properly. It is true that it will sometimes be found, even when the greatest care in feeding and attention is given, that a young sow, with a numerous small family, will suckle herself down in condition to such an extent that it is necessary to allow one or two periods of œstrum to pass before mating her ; or another plan frequently adopted successfully is to allow the pigs to remain on the yelt for a longer period than usual ; this often results in benefit to both sow and pigs, and the trouble, which sometimes follows a rest from breeding, of getting the sow to conceive when she is at last mated, is avoided. For some reasons, which are not far below the surface, a sow which has not been mated for two or three months after her litter of pigs are weaned, refuses to become in pig ; this is more generally the case when the sow has become fresh in condition. This difficulty has had the effect of making pigbreeders shy of adopting the plan of regulating the times of farrowing of their sows, since they hold that the loss is less from the young pigs coming at the time of year less suitable for little pigs, than often results from resting the sow and meeting with a difficulty in getting her to breed again. This shyness of breeding may be reduced considerably by feeding on good rather than on bulky food ; for instance, a few beans or peas will prove far more suitable food for a sow about to be put to the boar than a comparatively large quantity of sloppy and innutritious food. Nor is it fat that is wanted, as those sows which have become very fresh in condition during their rest will frequently prove the most troublesome. We have found that sows are far more likely to prove in pig when mated at the latter part of the period of œstrum than if put to the boar as soon as she comes in heat. If possible, the sow should be put in a place by herself for a day or so after her visit to the boar. Various plans have been recommended by which the difficulty in getting a sow to become pregnant can be overcome ; the most effectual plan, so far as our experience has gone, has been to mate one of this character as nearly as possible at the termination of her heat, and to have her served by several boars, the one after the other. Why this proves successful we do not pretend to state ; that it has been we know, and after other expedients have failed.

As soon as the owner has reason to believe that his sow
has conceived, and this can frequently be discovered by the
sow gaining flesh and showing general signs of well-doing,
it is advisable to begin to feed her on food of a somewhat
better or more nutritious and less bulky character than that
on which the barren sows are being fed, as the drain on
the system of the sow becomes greater, and one can easily
understand that a pregnant sow carrying fifteen to twenty
embryo pigs—a very common number in our own experience
—requires considerable support in the shape of muscle-forming
food. Many sows are unfairly treated during the latter period
of their times; flesh-forming foods are offered to her, and
her condition, so far as fat is concerned, appears to be satisfactory to the owner; this mere fat rapidly wastes as the
suckers require more milk, and the sow frequently fails to
rear her large litter successfully. From the sixth to the
fifteenth week of the sow's pregnancy she should be fed
liberally on those kinds of food which are best suited for the
production of lean meat and muscle rather than fat, as the
drain on the sow's system in building up the framework of
some fifteen little pigs is of a very extensive nature, and far
greater than it is considered to be by many people who have
not given the matter serious consideration. As the time for
the sow to farrow approaches within some ten days, it is
advisable to have her placed each evening in the sty in which
she is to farrow; the food for the last week or so may be as
nearly as possible of a character similar to that on which she
will be fed for three or four weeks after she has farrowed.
As to the best kind of food for newly-farrowed sows, there is,
as in many other departments of pig-feeding, a great diversity of opinion. This may be due to the varying character
of the sows; for instance, it is an admitted fact that ordinary
sows, of no pretensions to breeding or high-feeding qualities
can, without any serious consequences to their little pigs, be
fed on richer and more nutritious food than pure-bred sows of
a good strain. Indeed in many cases they require to be so fed.

If barley meal, beans, or similar food were given to the
pure-bred sow when she had young pigs on her the probability is that the youngsters would very speedily suffer from
indigestion, diarrhœa, or fits, whilst the pigs on the ordinary
sow under this treatment would make good progress. This
may be due to several causes, of which two may be the extra
richness or quantity of the milk given by a pure-bred sow
when fed on forcing food, or that the little sucking pigs from
pure-bred parents actually require less milk to make the
necessary growth, or that they are by their breeding enabled
to extract a greater amount of goodness from the milk of
their dam. For at least a quarter of a century we have
ceased to give barley or other meal to our suckling sows
. until the pigs are at least a month old. Our newly-farrowed
sows are fed on sharps, or what is locally termed thirds,
randan, dan, middlings, &c., and a varying amount of bran,
determined by the richness of the sharps, the number of
the litter of pigs on the sow, and the age and condition of
the sow. Then, as the pigs become four or five weeks old,
a little meal composed of wheat, barley, maize, oats, &c.,
is added to the sharps and bran, and the times of feeding
increased from two to three per diem.

It is scarcely necessary to remind our readers that a somewhat different system of feeding the sows is advisable in the
winter to that which is suitable in the summer, when there is
plenty of grass. It may perhaps meet the case of most pig-
keepers if we relate our own system of management of sows
from time of weaning to farrowing. At the time of writing
(December) we have some sixty aged sows, the majority of
which are carrying their pigs; some have been served about
two months, others as recently as a fortnight. These are being
kept in three lots, one of which comprises thirty-six of the
strongest and most lusty of the sows; these have the run of
some fifteen acres of grass, and besides what they can find on
the grass field they have nothing but kohl rabi, and an occasional feed of small or diseased potatoes. For shelter they

have the run of two yards, in one of which the cows are kept and in the other a number of young stock are fed on roots, straw, chaff, and a little meal, so that the sows are not able to pick up much of a living in the yards. As those of the sows which are more forward in pig require more nutritious food, they will be drafted out and supplied with it. Another lot of nine sows, which have each reared one good large litter of pigs and are again forward in pig, have the run of a grass field of some five acres in extent, in which is an open shed, which is used by the sows for shelter. Their food consists of kohl rabi and some mixed meal, of barley, wheat, maize and peas, fed to them as slop, night and morning. The remainder of the sows, which include the oldest and the weakest, have the run of a pasture, with some thirty yelts which have lately been put to various boars; these are fed as the last lot, except that the kohl rabi are passed through a Gardner's cutter and placed in troughs, on which meal stirred with water is poured. This plan is adopted since the old sows and the yelts are unable to break up the hard kohl rabi, the teeth of the older sows being worn and often loose, whilst the yelts are generally about to cut their temporary incisors or front teeth when they are about ten months old; even with the strong and fully developed sows, we find some of them unable to scoop out or break up the hard kohl rabi; last season it was more particularly the case as the rabi were of enormous size and thicker in the skin than usual.

This system of feeding the sows will continue until about March, when the rabi will have lost much of their goodness, and the supply of them will be exhausted. Mangels will gradually take the place of the rabi, but in smaller quantities; this for two reasons—the supply is shorter, and so early in the season the mangels are of less feeding value than when thoroughly ripe later in the spring. By the middle of April we usually have a good supply of grass, so that roots are not much required for the sows, which then do best with the addition of a few beans, or even soaked maize, or, better still, maize

5

meal soaked for twenty-four hours. We grow but a very few swedes or white turnips, or these would take the place of the kohl rabi. At one time we grew a considerable quantity of cabbages for the pigs, but we found that these caused constipation, and were not at all suited for little pigs or for young boars which were kept confined in styes. Even kohl rabi require to be sparingly used for the younger pigs, or they will sometimes cause constipation, and this, if not removed, will frequently be followed by fever, more or less dangerous.

Where it is not possible to give the sows a grass run, great benefit will be derived from giving tares, lucerne, or other green food to them; numbers of breeding sows are mainly kept on these green foods in some districts. Our own experience, and that of many other pig-keepers, is not in favour of the use of prickly comfrey; the pigs are not particularly fond of it, and, unless a considerable addition of good food is made, they will grow big in the belly and narrow on the back, losing all muscle. It is true that a great quantity of weight per pole can be grown of it, but its feeding value is, in our opinion, very little, owing to the enormous amount of water in its composition. In this district we could doubtless grow very heavy crops of it, as it appears to be indigenous to the soil, since one variety of it grows wild; the stock will not, however, eat it unless very much pressed for food. One drawback to allowing the sows to lie in the yard with the cows is that very occasionally a sow, or even young pigs, will acquire the very bad habit of sucking the cows. We have known instances of cows being completely spoiled by the pigs sucking them and biting their teats; sore teats have resulted, and in two cases garget and the loss of the cow's quarter have ensued. The unfortunate habit is, perhaps, first acquired by the slight exudation of milk from the bag, when the cow is lying down; the pig scents the milk, and then proceeds to search for it. If the cow is naturally a quiet one and used to pigs, it does not resent the attentions of the pig, which soon helps itself to something more than the mere droppings of milk from the udder. We have seen a cow standing up and allowing a pig to suck it as it would a calf, but the habit is more frequently indulged in by the pig when the cow is lying down at rest.

It is scarcely necessary to caution those who have several in-pig sows running together to give them plenty of trough room; the natural propensity of sows when feeding is to continually run from trough to trough, in search of more or better food than is to be found in each separate trough. Sows are also very masterful and vicious with each other, so that it sometimes follows that if there are not sufficient troughs, two or more of the strongest of the sows will secure far more than their fair share of the food.

Now that we have, in the *Live Stock Journal* and other almanacs, year-books &c., those most useful breeder's tables, no excuse is available to those who neglect to be prepared for the farrowing of the sow. The variations in the time which a sow will carry her pigs are comparatively slight, and these are pretty well regulated by the age and condition of the sow; thus old and weakly sows and yelts will most frequently bring forth a day or two before the expiration of the sixteen weeks; sows in fair condition will generally farrow on the one hundred and twelfth day; whilst strong and vigorous sows will frequently go for a few days over time. It will be found that the average period from service to farrowing of sows is, as near as possible, sixteen weeks. The owner of the sow, knowing the date on which the sow is due to farrow, should have had the sow shut up for at least a week in the sty where she is to farrow. The approaching parturition is generally preceded by the enlargement of the vulva, the distension of the udder, and the giving way of the muscles on either side of the tail. As soon as the udder becomes smooth and heated, and milk can be drawn from the teats by the pressure of the thumb and forefinger, the arrival of the pigs may be confidently looked for within the next twelve hours, unless it be a first litter, when the rule will not hold good. Generally the sow will take an hour or two to prepare the nest in which she is to farrow. Only a little straw, and that short, should be allowed the sow, as, should the straw be long and the sow restless during farrowing, the little pigs are likely to be smothered unless they are taken away as they appear, first being rubbed with a cloth, put to the teat to get a taste of the milk, and then placed in a hamper in which some dry wheat straw has been placed. In cold weather it may be necessary to cover over the hamper with a sack or a rug. It is also a good plan to nearly fill the hamper with long and dry wheat straw, so that a kind of nest can be made, into which the little damp pig will huddle with its partners, and quickly become dry and warm. If by any means the little pigs get a chill, and turn cold, limp, and damp, a teaspoonful of gin will help to revive them, and a suck at the teat will complete the cure if the weather is not very severe. We have of late years had one or two winters when at times the cold was so intense that it was simply impossible to keep the young pigs sufficiently warm. An attendant who has had much experience will at once know when this is the case by the peculiar cry which a chilled pig invariably gives in the stage before it becomes quite helpless and semi-unconscious. Should the act of parturition be a very protracted one, it is advisable, in very cold weather especially, to place those pigs which are dry to the teat, lest the long-continued deprivation of their natural food should result in their becoming chilled and troublesome to get to suck. If one's sows are particu-larly prolific, it will generally be found that the number of pigs farrowed are more in number than the sow's teats; it then becomes necessary to decide as to the destination of these surplus pigs. We have frequently had these little ones reared by hand when the sow is a particular favourite, or when, as in January, we were anxious to have a number of young boars or sows to show as pens of breeding pigs at the summer exhibitions, but as a rule it is far less troublesome and more profitable to select the best of the pigs, or the boar or the sow pigs, whichever you may require, and to knock on the head all the pigs beyond the number it is thought the sow will rear properly. We have killed as many as forty little innocents within one week, as we consider it far better for a sow, both for herself and the pigs, to rear ten good ones than twelve or fourteen, several of which will be like the dilling, wreckling, parson's pig, &c., of the litter. Opinions as to the proper number of pigs a sow should rear will differ, but the season of the year and the breed of the sow should influence the decision. Personally, we hold to this view—that seven or eight for a yelt, and ten to twelve for a sow, are sufficient to enable the mother to rear them well, without placing an undue strain on the sow.

As a rule the placenta will come away shortly after the arrival of the last pig; a practised hand will usually be able to tell when the sow has finished farrowing by the totally different action of the sow and by the peculiar twitching of the muscles of the shoulder, &c.; the afterbirth should be removed at once, and placed out of reach or scent of the sow; the little pigs which are selected to be kept should be placed to the teats, and if assistance be required—and it frequently is, since the youngsters often struggle most vigorously—the attendant should place the weakly ones to the teats and hold them in position until they have tasted Nature's nectar, after which little trouble will be experienced, unless a stronger pig takes a fancy to the teat at which the more weakly pig is regaling itself. If this does occur, the disturbed pigling must be directed to another vacant teat, as it is impossible to cause any of the strongest pigs of the litter to take to the particular teats you may select for them; they will do this for them-selves, but when they have made their final selection they usually adhere to their choice; but they are by no means averse to a stolen draught at another teat if it be possible to obtain it through the absence or inattention of its rightful owner for the time being. An impression appears to be abroad that it is possible to select for the various pigs the

different teats from which it is advisable they should draw
their supplies of milk; our experience leads us to believe that
this is impossible, since well-nigh each little grunter appears
to take a particular fancy to a teat without any apparent
reason for the selection, except, perhaps, because a brother
or a sister pig wishes for it. As a rule, those pigs which
suck of the teats nearest to the fore legs of the sow thrive the
best, whether it be from the greater quantity or the better
quality of the milk, or from both causes has not been clearly
proved. In this, as in many other points connected with
sucking pigs, the rule has exceptions, as we have sometimes
known the pig which has selected one of the very hindmost
teats grow and thrive as well as the best of the litter. It
appears to be generally acknowledged that it frequently
happens that those of the sow's teats which the pigs of her
first litter have drawn regularly give the better supply of milk
at the subsequent litter. A case in point came within our
observation recently. A sow was rearing ten very nice pigs,
seven of which were decidedly larger and fresher in condition
than the other three; it was noticed that the seven pigs had
taken possession of the very teats which her former litter of
seven pigs had sucked. We have often noticed similar
instances, which tend to strengthen the arguments of those
breeders who hold that it is advisable to allow the sow with
her first litter to rear as many pigs as possible, so as to
develop her milking qualities.

It sometimes happens that, owing to accidents, or to a sow
having but a small litter or several dead pigs, the number
of pigs left alive is fewer than she is able to or ought to
rear. Under these circumstances we have often taken pigs
from another sow whose family was too large and put them
on to the sow with a small litter; if the foster pigs are a day
or two old very little difficulty arises, but sometimes when
these pigs are a week or more old they will not take kindly
to the foster if they can hear the grunting of their own dam,
until hunger compels them. In two instances we have reared

two successive litters of pigs from one sow ; the dams of the reared litters died during or shortly after pigging, and rather than lose all the pigs or rear them by hand we weaned other litters of pigs which were some six weeks old and placed a portion of the newly-born little pigs on the sows from which we had removed the others. Very little trouble was experienced, but we found that the supply of milk ran short almost too soon for the second family ; the deficiency was made up by feeding new milk to the suckers.

Should the sow carry her pigs beyond the usual period of gestation, it frequently happens that the piglings' teeth will have made an abnormal growth, and in some instances the teeth will have become discoloured to an extent which has led to the common saying that "Pigs born with black teeth never do well." These little teeth are often very long and very sharp, so that, when the little pigs attempt to suck, the teeth extend beyond the tongue of the pig and prick the inflamed and tender udder of the sow, giving her great pain, which frequently causes her to refuse to suckle the pigs, and sometimes she will attack the little ones with open mouth, when one grab from her powerful jaws seriously injures, if it does not at once kill the youngster. Unless immediate steps are taken to remove the cause of this trouble, the pigs soon die for want of food, and the sow's udder becomes distended with milk, and inflammation of it follows. The remedy is simple and easily applied by the attendant on the sow. He takes up each pig, tucks it under his left arm, opens its mouth with his left hand, and with his right hand and a small pair of pinchers he breaks off the erring teeth, and then places the pig to the sow ; then, by a little of both coaxing and scratching, the sow will turn on to her side; the little pigs, being unable to bite the udder and each other, will quickly relieve the distended udder of the sow and prove a source of pleasure to her instead of an irritant and a cause of pain. Sometimes the sow will become impatient on hearing the shrieks of her little pigs whilst the operation of dentistry is progressing ; if this does affect her it is best to take the little pigs into an adjoining place out of hearing of the sow.

After the newly-born pigs have sucked out the sow, a little warm slop should be given to her; if her bowels are at all constipated or if she has had a hard or prolonged time of it, two ounces of sulphur and a nip of nitre may with advantage be given in the food. Whilst the sow is eating this the sty should be swept out, but the nest in which the sow has farrowed should not be disturbed, or she may be some time before she settles down again, as she will probably set to work to re-make it. If it be left intact the sow will most likely at once lie down, when the pigs may be put to her, and the whole party will rest comfortably until the morning and feeding-time comes. Before the sow is fed she should be walked out, when she will relieve her bowels and bladder; the latter is very necessary, as the operation is frequently a somewhat painful one after the sow has had a bad time of farrowing. The sow should be fed sparingly for a few days, after which we invariably feed our sows on sharps and a proportion of bran, varying from one-third to one-fifth, according to the quality of the sharps and the number of pigs on the sow; this feeding, as before stated, is continued for some four weeks.

It will generally be found to be best to allow the sow to get rid of her pigs unaided if possible. Should she be an exceptionally long time in starting, or should a very long time intervene after the arrival of a portion of the litter, and the sow show by her continuing to pain that other pigs are present which she is unable to bring forward, the attendant should smear his hand thoroughly with carbolized oil or lard and then insert it carefully, as frequently a pig will present itself crosswise, or the sow will only be able to get the head forward; in the former case the pig should be gently pressed into the womb, and in the latter the head of the little pig should be grasped, and gentle force be applied to bring it forward at each throe or pain of the sow. At times the little pigs will appear to have but little life in them when they are a long time coming into the world ; the attendant should then open the little pig's mouth, and blow lustily down its throat, so that the lungs become expanded, when the youthful grunter will quickly recover.

# CHAPTER VI.

## MATING.

On this particular subject there is perhaps as great a diversity of opinion as on any one of the hundred and one other points in connection with our subject.

This variety of opinion is, to a certain extent, due to the varying conditions under which the pigs are kept, and to the various systems of management adopted even by persons in the same neighbourhood. These alone would be sufficient to account for the success or failure of early mating, or that which is by some authorities so termed. Thus you will find certain pig-keepers argue that a young boar or a young sow should not be more than eight months old when first mated, for the reason, amongst others, that they will be less likely to be fruitful if kept apart for a much longer period, and that with a yelt or young sow having a predisposition to lay on flesh, very great difficulty will often arise in getting her to become pregnant if not paired when young. On the other hand, the neighbour, who is credited with being equally as successful a breeder of pigs, will maintain that the young boar or sow should be at least ten months old, and if a year old so much the better, as by breeding from pigs of a younger age their development is retarded and their produce are less robust and thrifty.

Although these views appear to be diametrically opposed, yet both are very likely to be found correct. This is owing to the different system of raising and feeding adopted. The pig-keeper who finds that eight months is the better age at which to mate his young pigs is most likely what is called a liberal feeder, or one who always has his pigs in good condition, and this is not in every case alone due to an extra amount of food being fed to the pigs, but rather to frequent feeding, and to care in the selection of the pigs and the food, and the form in which the latter is given to the pigs. Under this system the procreative powers of the young boars and sows are equal to those of older pigs kept on a spare diet or one not so suitable, and the young sows are fully able to rear the litter of young pigs which, in the ordinary course, will arrive by the time that the dam has completed her first year.

The owner of the young pigs which are not so forward in condition nor so well matured, owing to the different manner of feeding, and, perhaps, want of careful selection, will be acting with an equal amount of judgment in not mating them until they are ten to twelve months old, as, if the young sows were to farrow their first litter before they were some fifteen months old, the chances would be in favour of a litter of small, weakly pigs, which would require a considerable amount of nursing and attention, and, even then, would not make the same growth and improvement as a litter from a stronger, or even an older sow in the same condition.

Besides this question of development there are several other points to be studied. The season of the year when the young sow is to farrow her first litter has a most important bearing; should the young family be due to arrive in May or June, much less care and attention would be necessary, and the sow might be younger and less robust than if she were timed to farrow in September, since she, as well as the young pigs, would, in the latter case, be scarcely in a fit condition to withstand the cold and damp autumnal weather. The critical period for young sows suckling, as well as for their pigs, is when the latter become some five or six weeks old. The strain of supporting a litter of pigs will, by that time, be apparent in the sow; she will have lost flesh and energy, and be far more anxious to keep away from her pigs than to rejoin them, if, for the benefit of a change for the sow and an opportunity of feeding the youngsters apart from their dam, she has been turned out of the sty. The lengthening nights and often sunless, foggy days appear to check the growth of young pigs farrowed in the early autumn to such an extent that only strong sows in good condition should be timed to farrow in this, one of the worst periods of all for litters of pigs to arrive. For this and other reasons it is better to save the young sows intended for breeders from the spring litters of pigs. This gives the owner an opportunity of either mating them in the following November, so that two litters can be reared from her during the next summer, or of a litter being taken from her in the summer months, and the second trip in the January or February following. There is much to be said in favour of early mating; the risk of having trouble to get the sow to breed is less, and if she should go over two or three times—as is by no means unusual—the time lost does not appear to be altogether wasted, as the litter of youngsters will then arrive by the time the young sow is some fourteen months old. Even if she should hold to the first service, and appear to be too low in condition after she has suckled her litter of pigs, a month's rest will benefit her to such an extent that, by the time her second litter is due, all signs of the strain on the system will have disappeared, whilst she will generally prove a more careful mother and a better suckler than if the period of her idleness had been enjoyed before instead of after her first litter. Great benefit may also be derived by both the young sow and her litter by not weaning the youngsters until they are nine or ten weeks old.

Where only one or two sows are kept it does not matter much at what time of the year the little pigs arrive, providing warm, dry places, and those free from draught, are prepared for the winter pigs; but when there are a number of sows, it will be found advisable to so mate the sows that the pigs shall arrive between January and September, unless an attempt be made to supply some of the sucking pigs which are still fashionable about Christmas time. This is not a very profitable practice, as the value of these four or five-weeks-old suckers does not, as a rule, exceed 7s. or 8s. each. There are many persons who are careless as to the mating of their sows, with the result that many litters of pigs will arrive in the damp, foggy days of November, when the youngsters will make but slow growth, and thus the owners are tempted to get rid of the trouble of them by converting them into Christmas sucking pigs. In this way is the supply generally greater than the demand, and the price realised for them an unprofitable one.

If the majority of pig-keepers were asked as to the various points they have in view when mating their sows, very few would be able to inform you, so little attention is given by the average pig-keeper to that which is considered by the breeder of pure-bred animals a most important point. It is true that in too many cases there is but little, if any, choice in this matter open to the owner of the sow; but even amongst those who have sows enough to make it answer their purpose to keep a boar sufficient attention is seldom given to this matter. If a pure-bred boar is bought, sentiment or first cost often decides the selection question. To take the trouble of a journey to inspect one or more herds of pure-bred pigs, and to select a young boar likely to counteract certain weak points possessed in common by one's sows is a plan not frequently adopted ; the more general plan, or the one stated to be so, is to write to one or two of those who advertise their pigs, and then purchase of the advertiser whose price for age is the lowest. The excuse frequently made for this strange proceeding is that, as all the pigs are eligible for entry in the Herd Book, and thus must be pedigree animals, a guinea or even half a guinea may thus be saved, and of course gained. A greater fallacy does not exist. In the first place, the conditions of entry may not be such as to ensure that even

the grand-dam of a boar shall be of the breed of which
the boar is supposed to be, whilst instances have been
known of pigs being sold as pedigree pigs of one breed,
when the grandsire, and in a few cases even the sire, has
been of a different breed. So long as it is possible to breed
in two years a pedigree pig of one breed from a sow of
a totally distinct variety, so long will the value of so-called
pedigree pigs from many herds be valueless for the breeding
of any distinct form or type of pure or cross-bred pigs. The
mere fact of a pig, or of any animal, being entered in a Herd
or Stud Book is of very little real value to the breeder unless
the conditions of entry, and the system pursued by the
breeder for a great number of years, are such as to ensure
that the points possessed by the animals have for a certain
number of generations been specially studied by the breeders
of its forbears. It is almost invariably found that each
breeder of note has some particular point or points on which
he sets a very high value, and to the cultivation of these
special points in his animals will he have given a large share
of his attention. It may not be always quite possible to clearly
define the particular formation which each master of the
breeder's art most admires, or those points which he thinks a
fairly perfect animal of the breed should possess; still the fact
remains that the animals which have been for generations
bred by any breeder of note do possess certain characteristics
not common to the breed. It is this which gives so much
value to the sire from a herd of old standing. The mere fact
of a male animal being bred in such a herd, and possessing these
special points desired, renders it almost a certainty that these
qualities will be reproduced to a greater or less extent in the
offspring. It may not be always possible to find a herd, stud,
or flock, sufficiently large as to be self-supporting, and here
comes in the great value of a carefully-recorded long pedigree,
since by this is one enabled to discover the breeding of the
various sires, and from learning the names of the breeders
of these sires a very fair estimate may be formed of their

value and also of their style and formation. It is, of course, desirable in the breeding of pure-bred stock that the sire should be an impressive one; but it is, if possible, still more necessary when one is attempting to breed animals of any defined type or formation from half-bred dams. As a rule the sire from a really good old-established herd is far more likely to be impressive than one from a recently-formed herd, however successful the owner may be in winning prizes at the various shows. The mere winning of prizes is a comparatively easy task, if capital enough be employed, whilst the number of breeders of any particular variety of stock who possess the perseverance and the ability to continue for a lengthened period the breeding of uniformly good animals of a distinct type is very limited.

There exists amongst breeders a diversity of opinion as to the respective points which the sire and the dam should possess, but it may be taken for granted that the majority of successful pig-breeders pay far more attention to the form and quality of the boar than of the sow with which he is mated; for this several good reasons exist, amongst others that it is better within the means of most persons to purchase one boar than several pure-bred sows, that a greater amount of benefit can be derived from one male than from several females, since the boar can be mated with a large number of sows, and also on account of the general belief that the pure-bred boar has far more influence on the form and quality of the half-bred sow than has the pure-bred sow on the pigs which she may farrow to a half or cross-bred boar.

It is considered by many persons that if they require pigs of a certain form, these can be obtained by mating a boar having several of the points fully developed, and the others totally absent, with a sow possessing wholly dissimilar qualities—the argument being that the strong points of the one will counterbalance those of the other, and thus a uniformly good animal will result. This might be anticipated with some degree of confidence if it were possible to find two

animals which had for an equally long period been bred to
these points, but with cross-bred animals there would be
little certainty about the results ; the produce might possess the
good, and even the bad qualities of both the parents, and by
chance a portion of the youngsters of a litter might be animals
of average merit.  If the boar and the sow were both of long
descent, and had each of them many progenitors which had
been bred for the respective points possessed by each of them,
then the probabilities might be in favour of the production
of a litter of youngsters having the points desired fully developed ; but by far the better plan is to select both the boar
and the sow as nearly as possible of the form and quality
sought in the produce.   The mating of two animals of
strongly diverse types does not often result in the super-
excellence or in the uniformity of the offspring.

## CHAPTER VII.

## MANAGEMENT OF THE YOUNG PIGS.'

THIS is one of the most important, and, to many persons, the most difficult of the departments connected with successful pig-keeping. There are more pigs lost or irretrievably ruined when they are first weaned than at any other period of their existence. The bonhams miss the milk of the mothers, unless they have been allowed to well-nigh wean themselves, and even then they often eat their food too ravenously, and this causes indigestion and flatulency; or they eat too much at one time, with similar results. Many persons also imagine that it is necessary to give the newly-weaned pigs food of a far more nutritious character than that on which they have been fed when with the sow, or even at the time when they are fed apart from their dam before they are weaned; and this separate feeding, already alluded to, plays a most important part in the rearing of strong healthy litters. This over-feeding often results in a feverish condition of the little pigs, which shows itself in the dry state of the skin, in a desire to drink any water that the pig may find in the low places of the floor, and any liquid food in preference to that of a more solid character; then diarrhœa sets in, and if the attack is very severe the pigling wastes away and soon dies.

On the first appearance of constipation—indicated by the fæces resembling black peas—which almost always precedes the extreme looseness of the bowels, a fever powder (of which there are several in the market, such as Willson's, &c.) should

6

be given to the pigs, a change in the diet made, and a small
quantity of food given to the pigs each time.   A freshly-cut
turf, or even a shovelful of mould, a few cinders or small
coal will be readily eaten, and will prove of great benefit to
the feverish pigs.   In the States a mixture of salt and wood-
ashes is highly spoken of, and generally placed so that the
pigs can eat as much of this as possible, but from the fact
that American pigs are fed largely, if not chiefly, on maize,
the wood ashes or something similar in character are much
more necessary than when, as in England, bran, pea meal or
some food of this character is usually given to the pigs in
larger or smaller quantities.

Another most troublesome complaint with young pigs is
rheumatism, or, as it is variously termed, cramps, going off
their legs, &c.   Some old pig-men persist in asserting that
it is in many cases more like rheumatic gout than ordinary
rheumatism, since they contend that over-feeding on food of
too rich a nature is frequently one of the causes of its appearance amongst a lot of young pigs.   It is undoubtedly a fact
that cramp, as we will call it for brevity's sake, often shows
itself when the food given to young pigs is of too heating a
nature and in too large quantities at one time.

Far more frequent causes of cramp are to be found in
draughty, damp, and low-lying piggeries.   One careful ob-
server has given it as his opinion that cramp in pigs and
kennel lameness in hounds are both the result of their sties
and kennels respectively being built on damp spots or where
the springs are close to the surface.   This may be so, as it is
well known that very few things are more detrimental to
young stock of all kinds than a damp lair, whether this
arises from the site or the improper floor of the building,
or want of care in giving the young animals a dry bed.   Pigs
especially suffer from this neglect, as they naturally huddle
into a heap and creep into the bedding in cold weather, and
should the litter be damp the pigs literally steam when disturbed from their nest.   Then if they are fed on a large

quantity of cold slop they are frequently seen to shake with cold, and return to the damp lair and huddle one on top of the other, in the vain attempt to find warmth and to warm up to natural heat the cold food with which they have filled themselves. Under such circumstances it cannot cause wonderment if the little pigs go wrong in their lungs, bowels, or legs. In the long winter nights it is often found beneficial to have the young pigs disturbed, say at 9 or 10 p.m., so that the bladder and bowels are relieved; this is said to prevent cramp.

One of the secrets of success in the rearing of weanling pigs is the frequent feeding in small quantities of food which should in winter, and even at any time when the weather is cold, be given to them milk-warm. The young pigs will return to their dry bed thoroughly satisfied and warm, their digestive organs will at once proceed to the digestion of the food, and none of the warmth of the body, which means a certain portion of the previous meal, will have to be expended in making the freshly-eaten food in such a condition that it can be utilised.

As to the kind or kinds of food on which it will be the most economical and the best to feed the young pigs, this will very much depend on the district or part of the United Kingdom in which their owners are resident. There does not appear to be any additional food on which the young pigs will thrive so well as skim-milk, unless, indeed, it be new milk, and this last is not infrequently used in the months of April and May, when butter is always cheap, and very frequently does not realise more than 6d. to 8d. per lb. in those districts where many milch cows are kept, and only a sparse population exists. Still, we think that fat of as much value to the suckers, and nearly as easily digested, can be purchased and used more cheaply than cream or butter fat in the feeding of pigs.

It is advisable to continue to feed the pigs after they are weaned as nearly as possible as they had been fed with the sow, and by adding a little skim-milk to the sharps or other wheat offals, oatmeal and wheatmeal, or whatever the sow and pigs had been fed with, the change in the food is comparatively of the slightest. As the pigs increase in age an addition of oatmeal—this is expensive—barley, and wheatmeal to the sharps will be beneficial; a few peas or whole wheat given in the middle of the day, instead of the slop food, will be enjoyed by the pigs, and answers well.

In the good old times when pigs were not considered fit to put up to fatten until they were nearly a year old, it used to be the fashion to give them a good start for the first two or three months, and then to gradually wean them from the finer to the coarser foods, and to convert the pigs into growing stores, in which most unprofitable state they continued to exist for many months, having to forage about the stack-yards, cattle sheds, and other places for at least a portion of their daily food, making but little growth and no progress from the consumption of a considerable quantity of inferior food, most of which had been utilised in keeping up the animal heat of the carcase, and in the furnishing of the necessary force for locomotion, which at various times, such as when trespassing in the garden, the barn, and even occasionally the kitchen of the farmhouse, required to be of express speed to avoid the various missiles hurled at their unfortunate carcases, or to escape the too impressive attentions of the general purpose dog on the farm to pendent ears or uncurled caudal appendages.

Although the march of education and common-sense has done much to change and to greatly improve on such a wasteful and unprofitable state of affairs, yet we occasionally see a few of these relics of the dark ages in the form of a long-legged, slouch-eared, lanky railsplitter, whose hang-dog look plainly betokens the life he has led, a portion of this having been spent in marauding expeditions, and the other in attempting to escape from his well-earned deserts. Pig-keepers have none too soon begun to realise that length of

life in a pig means food consumed, and that food costs
money, or, in other words, that time is money as applied to
the lengthened existence of a pig. Many are the farmers and
pig-keepers who now concentrate their energies to produce a
given weight of pork from each pig at the earliest possible
period, and to this system still more farmers and others must
eventually turn, as in it there is far more profit and consequently pleasure than in the old semi-starvation principle.
The young pigs, when two to three months old, are fed
three or four times per day, and as they grow stronger the
quantity of meal is gradually increased until many of the
pigs are sold off as porkers when four to five months old,
weighing from 65 to 75 lbs. each when dressed, and realise
some three to five farthings per lb. more than the carcase of
a fat pig weighing 150 lbs., and a still higher relative price
than a 350 lbs. pig.

Other pig feeders, who feed for the bacon-curers, often
allow their pigs to have a considerable amount of exercise
when they are from three to five months old, and then
proceed to force them along, mainly by the aid of meal, in
order to get them to present a carcase of pork of the weight
of about 140 lbs. by the time the pigs are seven or eight
months old. Some persons object to this loss of time and
food, and continue the liberal feeding during the whole of the
pig's short life, and turn out their pigs fat enough for the
bacon factory at the age of about six months. To us this
appears to be far the most profitable system, but to carry it
out successfully it is imperative that the pigs should be of
a quick-growing and strong-constitutioned variety. They
should be of one of those kinds that are not so prone to make
fat rather than muscle or lean meat, as are most of the small
varieties of pigs, particularly those which have for any length
of time been bred more with a view to showyard successes
than the production of a good commercial carcase of pork.

A paragraph which recently went the round of the press
was credited to a person whose experience with pigs had

been chiefly confined to fatting, slaughtering, and converting them into bacon. It was to the effect that to feed a suckling sow in the same place as that in which the little ones were located was sure to lead to trouble, inasmuch as the little pigs would try to eat some of the sow's food. This is exactly the very thing that a thoughtful pig-feeder likes to see, so that the strain on the sow may quickly become reduced and the youngsters less dependent on their dam.

Many persons adopt methods of various kinds to encourage the little pigs to eat as early in life as possible, and we may safely trust to nature to have given to young pigs, as to all other animals, just sufficient instinct as to know when their teeth and their digestive organs are sufficiently developed as to make the best use of some other than liquid food. We have frequently noticed that the best litters of pigs at weaning time are those which have taken most kindly to feeding in their early youth, and further that their dams are in better condition, showing less of the effects of that strain on the constitution inseparable from the charge of a large family of bonhams. One of the most successful of our pig-breeders of times long ago—Mr. Stearn—used to advocate the feeding of the little pigs at a very early age, by placing a trough in a place out of reach of the sow, in which was placed new milk, &c. It is of course necessary that care should be taken that the sow is fed on food which is easily digested, such as sharps, &c.; then the youngsters will take no harm in joining the sow at her meals. Should barley meal, grains, or similar food be given to the sow, as is recommended by some authorities, then it would be advisable not to allow the little pigs to have access to the trough, as such food would almost invariably lead to the little pigs suffering from indigestion, constipation, diarrhœa, and other complaints to which injudiciously fed pig-flesh is heir.

Sometimes the pigs of a small litter, or those belonging to a particularly good milking sow, will become too fat and affected with fits. These pigs, when attacked, suddenly fall over, remaining almost motionless for a few minutes, when they gradually recover, until after repeated attacks they die. The general cause of these fits appears to be over-feeding of both pigs and their dam ; the chief remedy, therefore, is to reduce the quality and the quantity of food, to give in the food a gentle aperient and also to place within the reach of both sow and pigs a lump of earth and some cinders or small coal.

When the piglings are some three or four weeks old the sow should be turned out from the pigs for an hour or two in the morning, and during her absence a small quantity of sharps stirred, if possible, with skim milk, is fed to the youngsters, who will readily eat all they wish for, whilst the remainder, if any, will be eaten up by the sow on her return into the sty. Should there be a paddock or grass field con-venient and no difficulty be experienced, both sow and pigs will thrive better in the summer-time if the sow be let away from her pigs twice or even more frequently each day. The sow will then obtain the greater part of her food in the form of grass, whilst the sties are kept cooler and sweeter for the young pigs. A few peas or whole wheat will be much relished by the little pigs in the absence of the sow, and if sparingly given will result in benefit. The wheat is perhaps better for the little pigs in the warm weather and the peas when it is colder, although we like to give them both at times ; little pigs, like all young animals, thrive best on a diversity and change of food.

When the pigs reach the age of five or six weeks they will eat a considerable quantity of food and thus become far less dependent on the sow, whose terms of absence may be ex-tended, until at the age of seven to eight weeks in summer, and nine to ten weeks in winter, the pigs will become gradually weaned. The pigs will not then miss the mother, nor the sow be likely to suffer from an over supply of milk. Those of the young pigs not intended to be kept for breeding pur-poses should be attended to when they are about six weeks

old, care being taken that they be not fed for at least twelve
hours before they are operated upon, and a limited quantity
of food be given to them for two or three days after.  There
will be far less risk and suffering in this early attention than
if the pigs were left longer, or until after they were weaned.
In years gone by the castrators would not usually undertake
the castration of a young male pig which was ruptured, or,
as it is sometimes called, blown.  Now it is not considered at
all a difficult operation, one incision only is made in the
scrotum, and this incision sewn up after the testicles are
separated from the pig.

One of the secrets of success in raising young pigs is to
feed often and little at a time.  Anyone not thoroughly
acquainted with pigs would be surprised at the very great
number of times during the twenty-four hours a sow suckles
her little pigs.  For this there appears to be at least two
good and sufficient reasons; one is that the sow is unable to
carry a large quantity of milk for her numerous family, and
the other that the stomach of the pigling is not capacious
enough to stow away any great quantity of food at one time.

As the pigs arrive at weaning-time we generally mix with
the sharps a small proportion of meal; this we increase as
they grow older until at, say, ten weeks old the youngsters
will thrive well on one-fourth meal and three-fourths sharps.

One most important point in connection with the management of young pigs is the general belief amongst buyers of
fat pigs that the meat from a pig which had been kept in a
progressive state from its youth up is of a finer quality than
the meat from the carcase of a pig which had been allowed,
as is far too frequently the case, to live a considerable time as
a store pig, in the hope—vain though it be in the majority of
instances—that the growing pig will eat and pay for certain
odds and ends about the farmyard that would otherwise have
been wasted.  The bacon-curers are stated to place a higher
value per stone on those fat pigs which are fattened early in
life on the concentrated system of feeding, *i.e.*, in such a

manner that they never receive a check, and are, therefore, fat at much earlier age than the ordinary pig of the country. The breeders and feeders of pigs must continually bear in mind that the only hope of making pig-keeping really profitable lies in their producing pork of the very best quality, and this can only be accomplished by alone breeding from those animals which are possessed in a marked degree of those points such as early maturity, quick growth, fine quality of bone and offal; and then by so feeding the animals that every advantage is taken of the two former essential qualities in their feeding stock.

## CHAPTER VIII.

### EXHIBITION PIGS.

It has, at times, been confidently asserted that the exhibition of pigs at our agricultural shows has not proved of benefit to breeders of pigs generally, owing to the major part of the prizes having been awarded to animals—especially in the classes for older pigs—totally unable to produce healthy offspring, and to so many of the prize-winners not being of the form and type demanded by the purveyors of pork and the curers of bacon.

Although we cannot admit the entire truth of this conten-tion, there undoubtedly has been much cause for complaint, both in the classification, in the qualifying clauses, and in the decisions at many of the minor, and even at times at the larger shows. In the classification, an almost entire absence of originality is evident amongst the councils or committees of the societies responsible for the prize schedules. For years the same old classes for boars, for sows, and for pens of yelts held good at well-nigh all the county shows, so that often the itinerant exhibitor simply bought two or three old fatted boars, the same number of sows, and a pen or two of fatted yelts for his showing stock-in-trade, with which he could spend some two or three months from home travelling from one show to another and killing all healthy and local com-petition, since it would not pay anyone to show his stock in fair breeding condition against the over-fatted pigs of the

MIDDLE WHITE PIGS.

Winners at R.A.S.E. and other Shows.    Bred by and the property of Mr. A. C. Twentyman.

professional exhibitor, nor would it answer the purpose of a resident in the county to feed up his animals to such a pitch as to risk their usefulness as breeding animals to win one or two prizes in the season. To such an attenuated condition had the pig department arrived at some of the county shows that various alterations were made in the conditions of showing and in the prize schedule. These were for a time successful in obtaining a greater number of entries, but the bad times, or that which it has become common to term the agricultural depression, began at about the same period, so that many farmers gave up showing their pigs. The somewhat reduced numbers of subscribers to the county societies was also made use of as an excuse to reduce the value of the prizes offered for pigs, whilst the charges for railway carriage, for cartage, and for attendants, became heavier rather than lighter, so that even the owners of the prize-winners found that little, if any, direct profit resulted from the exhibition of their pigs. In fact, it was patent that only those owners having long purses wherewith to buy the best animals of the year that could be procured could, under the conditions prevalent, continue to exhibit pigs, and then only by sending to a number of shows and thus securing the best of all advertisements at that period. Even in this a considerable change appears to have been gradually taking place, owing to the public not having quite so blind a belief in the theory that because an exhibitor wins prizes, his herd is of necessity composed of really first-rate animals.

The suggestion has been frequently made and warmly supported, in some quarters, that the animals shown at the county shows should either be the property of the exhibitor resident in the county or be bred therein. This idea has much to recommend it, as a direct and strong incentive would thus be furnished to farmers and breeders of stock to send their animals to the show in fair condition, with an assurance that they would not have to compete with stock belonging to wealthy outsiders, whose exhibits, bought at high prices, are frequently so loaded with fat that, to use a vulgar but common term in the showyards, they "smother" the animals shown in breeding condition.

If these restrictions were enforced at the county shows, these latter would be far more likely to carry out the original intentions of the founders of local or county shows, *i.e.*, the encouragement of residents in the districts to improve the farm animals and to exhibit the best of their own stock, to act as an incentive to their less enterprising neighbours as well as to create a generous rivalry amongst the farmers in a county, and even sometimes amongst those occupiers of the farms on adjoining estates. We are strongly of opinion that infinitely more interest would be generally taken by the residents in a county if the competing animals were the property of their friends and neighbours. In times gone by, before shows were so numerous or so extensive as at present, a winner at the "Royal" show was looked upon in somewhat the same light as the fat woman at the fair—an object of wonderment. Now this is all changed, and few country people of middle age exist who have not been to the "Royal" or to the Smithfield Club Shows, whilst our stock and agricultural papers are so cheap and so profusely illustrated that a knowledge—although sometimes rather a faint one—of the style and form of the chief winners can be easily acquired by any one at a very moderate outlay.

The classification at some of our summer shows might be still further improved by reducing the limit of the age of the competing boars. At the present time, old boars appear year after year at the same shows, many of them being kept for no other purpose than to be exhibited, and thus competition is stifled, as a young stock boar in a breeding condition has but little chance in competition with a well-known prize-winner shown in what is termed perfect show form. Exactly the same tactics are pursued with the old bloated dowagers whose maternal duties have long been in abeyance; these useless old fat monsters take the prizes away from a really good sow in a breeding state, and even at those shows where it is necessary that the prize sow shall qualify by producing live pigs within a stated period this regulation is not so carefully or thoroughly enforced as it should be; besides this, the small amount offered as prizes is not so much the object with many exhibitors as is the fact of owning the animal to which the prize award is made, and the having it recorded in the Press that from the owner's herd the winner had been sent to the show. The advertisement of one's herd is with many exhibitors one of the chief points aimed at.

If the age of the boars exhibited was limited to two years, far greater competition would be engendered, and the boars which had proved themselves to be the best of the various breeds could then be utilised for breeding purposes without the risk of their reproductive powers being seriously affected, as they must be when the boars are kept up in show form for four or five years. There cannot exist a doubt that some of our breeds of stock have been much injured by this show fever; many of the best animals—especially females—have been spoilt in the training, and the services of most of the old show animals have been lost to the country. If the sows were obliged to be shown in milk, and their litters exhibited with them, we should be much more likely to have the best brood sows honoured, and many of the sows which have in the past won most of the prizes would either not have been shown or would have had their inferior characters as breeding sows thoroughly exposed. The form sought for in the show sow by many judges is exactly the opposite of that which a really good suckler would possess, and yet this ability to give a good flow of milk for its numerous progeny is one of the chief essentials in a brood sow.

An alteration might also be advantageously made in the classification of pigs at our fat stock shows. If the attempt to award sets of prizes to all the various so-called pure breeds were to give place to an endeavour to honour those pigs which reached a certain standard of length, weight, and quality of meat in the least time, the public would then have

an object-lesson as to which of the breeds of pigs are com-
mercially the best. This alteration might necessitate other
changes which appear to be most likely to prove beneficial ;
one of these would be that the pigs at our fat stock shows
would have to be weighed, and thus onlookers could at once
form an opinion as to the value of the pure or cross-breed of
the pig exhibited to convert meal, &c., into meat, and the
present system of guessing by the judges as to the weight of
the pigs exhibited would be done away with, to the great
benefit of some of those exhibitors whose pigs, at the fat
stock shows, have not in the past received their full share of
the honours. But very few of the gentlemen who are in the
habit of acting as judges of fat pigs are able to form a fairly
accurate estimate of the weight of a fat pig. This may
appear to be a strong assertion to make, but we have not the
slightest doubt that it would be completely proved by the
introduction of a guessing competition—somewhat after the
style of those recently so popular in connection with fat cattle ;
but we would vary this slightly by placing before the competitors a pen of pigs of the Berkshire, of the Large, of the
Middle, of the Small Whites, and of other breeds. The variation in the guesses made would surprise most people. Another
of the changes perhaps requisite with the new classification
might be the appointment of a pork butcher, or the buyer for
a bacon-curing firm, or someone of that kind, as one of the
judges. It is doubtless a fact that some of the prize-winners
in olden times would have fared badly with such a tribunal ;
still the shows would be of infinitely greater educational
value to both breeders and exhibitors of pigs.

It would be advisable to have classes for pigs of any pure
breed, for pigs of any first and distinct cross, and mayhap for
nondescripts, or pigs of no particular variety or cross, and
then a champion prize for the best pen of pigs in any of the
three sections. We would also suggest that separate classes
should be provided for pigs under four months old and not
exceeding 100 lbs. live weight, for pigs not exceeding six

BERKSHIRE PIGS.

Winners of Champion Prize at Smithfield Club Show.        Exhibited by Mr. John Newton.

months of age, and for pigs not over nine months old. The class which at present exists for pigs between nine and twelve months old would be expunged, as every pig kept for fatting purposes, after it is eight to nine months is a direct loss to the owner and an indirect loss to the country, since a pig of that age ought to have been converted into meat, as the cost of keeping it in store condition is an unprofitable outlay, and the attempt to further fatten a pig after it weighs 300 lbs. necessitates a far greater expenditure of food than is profitable, compared with the manufacture of pork with a pig weighing 100 to 200 lbs. With many of the fat pigs shown, the last three or four months of their existence is a dead loss, as the food eaten by these pigs gives a comparatively small return of meat, and the extra weight gained is of little value per lb., whilst the whole carcase is of a considerably reduced value to the consumer, and to the butcher. In years gone by it was thought desirable by some breeders of pigs to have those monster animals, in order to prove to the public and to the breeders of ordinary pigs the great capacity possessed by the pure-bred stock to grow and feed to great weight ; but the need for this no longer exists—if it ever did exist — as the pig now required for breeder, feeder, and consumer is one which will convert the largest quantity of food into the best carcase of pork within the shortest time, or, in other words, at the present time a machine is needed for the speedy and economical manufacture of pork of high quality, since it has been most clearly proved that pork made with young pigs is far more cheaply made, that it realises the highest price in the market, and that it enters much more readily into consumption.

It matters not what the British farmer manufactures, he will in the future, far more than in the past, find it is imperative that the article he produces shall be of the best quality, and that every endeavour shall be made to reduce the cost of production, and we would venture to assert that in few of the articles which he manufactures is there greater

room for improvement than in the general run of fat pigs.  It
may be thought desirable that, owing to long-continued and
somewhat extensive experience as an exhibitor and a judge in
the showyard, some of the points required in the parents of
exhibition pigs should be touched upon and a few of the trade
secrets should be revealed.   The same desire was experienced
some thirty years since, when pedigree pig-breeding first
engaged our own attention ; but little or nothing was at that
period to be obtained from books or other publications, and
the state of affairs is not much changed at the present time.
One may read everything, practical or the reverse—and if
he reads everything written about pig-keeping he will find
much of the reverse—and then, unless he has a natural aptitude in mating, or an intuitive knowledge how to select and
mate the breeding-pigs in his herd, his showyard success will
not be great, or at all events lasting.   It might be considered
the height of absurdity to paraphrase the saying that " Poets
are born, not made," and to apply this to the breeding and
exhibition of pigs, but it is none the less true that the art of
breeding the finest of our various breeds of stock is a natural
gift.   But this alone would prove of little value ; it must be
allied with and aided by perseverance, a determination not to
be dismayed by failures, a keen perception of small peculiari-
ties in the various animals, and, perhaps of more importance
than all, constant personal attention.   We would not for one
moment assert that all these qualities are absolutely necessary
to gain a few prizes at our shows, but to gain a real and
continued success as a breeder and exhibitor of stock these
qualifications are undoubtedly more or less requisite.

Then with some readers the question will arise—How are
these qualities to be utilised, and what line did you take ?
This is a question much more easily asked than answered, as
the old hand scarcely knows how or when the first successful
start was made.   At all events, the tyro may take it for
granted that his chance of success will be far greater if he
purchases for his foundation stock boars and sows which have

LARGE WHITE PIGS.

Winners of First Prize and Breed Cup at Smithfield Club Show.    Exhibited by Mr. F. A. Walker Jones.

for as many generations as possible been bred from prize-winning animals. It does not necessarily follow that all of the descendants of these well-bred animals will prove to be first-class, but there is an infinitely greater chance of their being superior specimens of the respective breeds than if short-pedigreed and prize-winning "come-by-chances" were bought for the founding of a herd. A careful study of the breeding of the major part of our show animals will furnish the strongest possible evidence of the correctness of this view. We have noticed studs, herds, and flocks suddenly jump into notoriety by the purchase of a few apparently first-rate animals of short or of doubtful pedigree, and picked up from various breeders in different parts of the country ; or the sudden and somewhat brilliant success will be due to the lucky production of a few short-pedigreed animals in a large collection of breeding animals on a farm ; but the success will be as evanescent as brilliant. The far and widely collected animals, which will, perhaps, win many prizes, will prove to be utter failures when mated together for the purpose of breeding show animals, and the same result will follow the next crossing of the short-pedigreed animals, the cause in many of these cases being exactly the same, the prize-winning animals being to a great extent cross-breds, and the good animals being those having the outward semblance of one or other of the pure-bred sires or dams.

It may not be generally admitted amongst purists of breed-ing, still it is a fact that the produce of a sire and dam of two distinct pure breeds is often a better animal to look at and for slaughter than either its sire or dam. The cross-bred will often be more vigorous, of quicker and of larger growth, and thicker fleshed, but the experiment must not be carried further, especially by using the male produce, as there will be little certainty as to the form and quality of his stock. The first-cross female may be a most useful dam, providing a pure-bred sire, and one from the same source, is used. Here, again, comes up a point of importance to the beginner

7

who is hoping to found a herd good enough to take a high position ; it is this, to select the stock bull, ram, or boar from the same herd for a lengthened period. It has been the practise with some considerable number of new beginners to select a really good sire from one old breeder ; then, to secure, as they think, the good points for which the successful stock of another old breeder is noted, they proceed to buy a sire from the second herd, and so on. In many cases this will end in partial failure, the stock bred will lack uniformity, certain points will be very fully developed, and strong failings equally noticeable. In fact, the results will often be similar to those which follow the mating of cross-breds, and for this reason—that the various successful herds have been bred on totally different lines, and although the animals from each are perfectly pure in breeding or pedigree, and first-rate specimens, as a whole, of the respective breed, yet in form, style, and character, the animals in each of the herds may be so diverse in type and appearance that the first mixing of the blood may be almost the same as the crossing of two pure breeds. It is true that this divergence from the best style or type of the pure-bred animal will be lost much sooner than it would be in the breeding from cross-breds, and tolerably uniform animals appear after the first or second generation, provided the sire and dams produced in the newly-formed herd be mated ; but the difficulty will most likely be intensified if a third herd be requisitioned to supply a sire to be used in the herd. On the particular question of the selection of foundation stock separate chapters are given.

As to the best means to adopt to so bring one's pigs into the showyard in such a form as to deserve success, these are very simple and few in number. The main difficulty is to breed the pigs good enough in form and character, and having constitutions so vigorous as to be able to withstand the forcing requisite to get them into show form at an early or later age, according to the age at which they are intended to be exhibited. A point here to be considered is the early maturing qualities of some families or tribes; these must be utilised for the classes for younger pigs, and those tribes or strains which require more time to furnish and develop may be reserved for providing pigs for exhibition in the classes for more matured animals. There are also a few, but a very few, tribes of pigs that will furnish animals which will not only be successful in the classes for younger pigs, but also train on and continue to take prizes for some years. Of course these are the kind of pigs which should be secured by the novice, no matter what the cost may be, as the qualities they possess are not only most valuable for exhibition, but for breeding purposes. The same rule holds good with pigs as with most other kinds of stock, viz., that the very great majority of the most successful show animals of a particular breed will be found to trace to a few particular families or offshoots of it. It is true that an occasional winner will appear from within the ranks of other families, but the particular points of quality, style, action, compactness of form, and of others which it is impossible to define, but which are at once apparent to a really good judge, will not be observable in the produce of this come-by-chance prize-winner. In some cases, even the successful young animal of this character will mature into quite a different style of animal; its true character will come to the surface, and as it grows older the more apparent will its weaknesses become, whereas with an animal which has been for generations bred on the right lines, and from first-class parents, a marked improvement will for years be observable, with the result that an animal which in its earlier days might be considered just a fairly good specimen would in the end furnish into a really grand representative of the breed.

These last are the animals which the novice should attempt to purchase for his foundation-stock, as they must be the possessors of good constitutions, of soundness, of fine quality of bone, of legs and feet which were intended to carry the carcase with celerity and safety, of quiet dispositions, and of the hundred and one other good qualities which go to make an animal as near perfection as it is possible to breed one.

Having "caught your hare," or bred the litters of pigs from
which you confidently anticipate the pleasure of selecting a
goodly number of prize-winners, the beginner must beware of
the far too common plan of beginning to force the youngsters
as soon as they will eat ; by so doing the probable chances of
success are frequently complctely disposed of, as the youngsters
will become feverish and often receive such a chcck, even if
they live, as will effectually prevent thcir being trained for the
showyard to be exhibited in thc classes for young pigs. It is
far better to try and forget that in these litters are embodied
your hopes of future success and notoriety ; lct the sow be
fed in the ordinary way, and thcn, when the pigs are some six
or seven weeks, feed both sow and pigs on rather richer food
than you would if the youngsters were intcnded for store pigs
only. If the weather is at all genial, let them have an occasional run for a quarter of an hour, when the sun is not very
bright and hot. The sow may also continue to be fed somewhat more generously, so that the milk-flow will be larger and
continue for a longer period, in order that the pigs may remain
on her until they are ten to twelve weeks old. In this, as in
most other things connected with the training of pigs, a variation may with advantage bc occasionally made ; sometimes the
youngsters will actually thrive better when weaned than they
have previously thriven whilst suckling. The necessity for
weaning them will show itself in the coats and skins of the
little pigs not looking bright and shining, the great objection
of the sow to return to the pigs after being lct out of the sty
for exercise, and the evident reluctance of the pigs to suck the
sow. The after-treatment of the youngsters may be on much
the same lines as is recommended in the chapter on the
management of young pigs, only, if possible, let just a little
more of that personal care and attention be bestowed on the
coming show-pigs ; the extra food will necessitate an equal
extra amount of exercise, and mayhap at times a gentle dose of
medicine, which may consist of some simple remedy such as
is given in the chapter on the diseases of swine. One of the
great secrets of successful training of show-pigs is to give them

exercise frequently, not necessarily for a great length of time, but enough to give them an opportunity to stretch their legs, to ease their bladder and bowels, and to pick up a little grass, earth, &c., which will act as a tonic and as a medicine.

One most important point should be kept in remembrance, and most strongly impressed on the mind of the feeder—that the young boars and sows are intended for breeding purposes. We are aware that the temptation to give to the growing pigs various foods and mixtures, which will enable the pig-exhibitor to bring his pigs in apparently fine show form to one or two exhibitions is great, but the results are disastrous to the owner if he wishes to go the round of the shows with the pigs, and to the purchaser who may be tempted by the sleek and fat appearance of the pigs, and to the fact of a prize having been won, to buy one or more of them with a view to keeping them for breeding purposes. We have frequently had complaints made to us that the show pigs which were bought of certain exhibitors almost invariably proved disappointing; they ceased to make the progress which they appeared to have made, and very frequently the litters of pigs—if any—from them would be small, weakly and totally unlike the sire or dam when these were being successfully exhibited. There is little doubt that frequent experiences of this kind have led to the opinion which is most general, that the fitting up of young pigs necessarily spoils them for breeding purposes. This does not by any means follow, providing the young pigs have been trained and fed with care and judgment, rather than with the sole view of winning a prize at one or two shows, heedless of the harm which is certain to result to the young pigs exhibited. Self-interest alone, if nothing else, should, but doubtless will not, put a stop to this suicidal practice, as it is bound to ultimately result in a loss to the exhibitor who pursues it. His customers will not only desert him, but they will warn their friends against buying any of his stock, and should he fail to sell the pigs so unnaturally forced and fed, he will find them almost useless for breeding purposes, and thus his herd will gradually disimprove owing to his having spoilt his best pigs in his over-anxiety to win prizes. One of the great secrets of the success in the exhibition of pigs is to so feed and train them that their breeding qualities are not impaired. That this is possible we have no hesitation in asserting after an experience of something like thirty years, and on a scale of some magnitude, as will be evident when we state that in one summer season as many as one hundred and nine prizes have been won in the Royal and chief showyards by pigs bred in our herd. To be able to do this the new beginner must purchase the very best of stock from a herd which has proved to be capable of furnishing for several years exhibits which can, without undue forcing, not only win prizes at one or two shows during a season, but will also stand the strain resulting from the continuous showing during the summer. We have seen pigs of all ages begin the season in May, and after being exhibited at various shows all the summer, the older ones have come out nearly as fresh in August as when they started, whilst the young ones will have grown, gained flesh and improved generally. To succeed in this way the exhibitor must have the proper material to start with, and he must have *trained* the flesh on to his pigs ; the mere getting them fat is of no avail if several shows are to be successfully attended, and the exhibits are to be of any use for breeding purposes after the show season. Time and exercise are also necessaries for the successful exhibitor. All those various foods or condiments, such as sugar, &c., which merely add to the bulk or fat of the pigs, must be carefully avoided ; muscle and lean meat must be walked on to the pigs, and the latter fed on the best and simplest food possible. A mixed diet is of course necessary— sharps and wheat, barley, oats, peas, &c., all as finely ground as possible, will be found the best groundwork of the food, whilst mangels, tares, lucerne, clover, &c., will materially help to keep the show pigs in health. We have observed that cabbage has been recommended for the feeding of pigs ; our experience of this is not favourable, especially for show pigs, as this and kohl rabi to a slighter extent cause constipation.

# CHAPTER IX.

## BREEDING CROSS-BREDS.

The various pure-bred and local varieties of pigs afford ample scope for the efforts of those who are fond of experimenting and of those who consider that a cross-bred pig is generally more profitable, for ordinary purposes than a pure-bred one. In the United States, where the pig population has at times reached some 53,000,000, but which at present numbers some 8,000,000 fewer, several attempts—and successful ones too—have been made to originate a variety of pigs dissimilar in some unimportant respects to the pigs of certain breeds which are kept in the British Isles; but it is a curious circumstance that as time progresses these apparently new American breeds gradually become less dissimilar to the kinds which have been cultivated at home. The so-called Miami, or Poland China breed, affords a striking instance of this, as the pigs of the breed appear to be undergoing very similar changes in colour and character to those through which our Berkshire pig has undergone within the memory of some of our oldest pig-breeders. It appears to be probable that there will always exist one small and unimportant difference in the form and hang of the ear, which in the first instance was due to the use of a so-called Bedfordshire pig in the foundation of the Poland China. These Bedfordshire pigs were said to have been imported from the English county of Bedford some fifty or more years since, and from this fact acquired their name, but

from the description given of them it is most probable that
they were very similar to, if not identical with, the large, coarse,
lop-eared pigs common at that period in the Ouse valley and
in the Fens. Even within the last quarter of a century the
great majority of pigs in these lowlands were of a somewhat
similar character; indeed, it is by no means difficult to find many
of them now in those parts of the Fens far removed from the
railways or the beneficent influence of a good herd of pure-bred
pigs. As with pigs so with cattle; the effect of a good boar or
a good bull on the local stock is simply marvellous. The
results from the use of a really good boar are, of course, far
greater and much more quickly observable, since its produce
will be more numerous and arrive at maturity much quicker.

This peculiar formation of the ear of the Poland China pig
will continue, at least so long as the breeders of it set so high
a value on the peculiarity as is the case at present; but with
this exception, and perhaps that greater proportion of muscle
resulting from crossing and which has not yet been bred out
by in-breeding or other attempts to obtain that unduly highly-
valued property which is commonly called quality, there
appears to be but very little difference between the present
Poland China and the English Berkshire of some twenty
years since.

It is true that the breeders of Poland Chinas are more
favourably situated for the preservation of that natural muscle
and lean meat, which is frequently lost by want of care or
knowledge of the art of breeding, or the inability to obtain
from outside sources really good stock pigs, so that in-breeding
is not closely followed. Not only do the Poland China herds
in the States number some thousands, but the area over which
they are found is so enormous, and the variation in climate so
great, that the effect of purchasing a boar from another
American herd may be as great as the purchasing of a boar
from an English herd for use in the States. Until within the
last few years this particular effect of change of climate and
conditions of food and existence on animals does not appear

to have received due recognition. It is now held that the reintroduction into a herd of a male animal of exactly similar breeding, but which is descended from stock which has for a few generations been reared or kept in a different climate, is equally as beneficial as the use of a sire of a different strain of blood. If this be so, and it accords with our experience, another great advantage accrues, since there the great risk is avoided of losing the particular type for which the herd may have acquired notoriety, or be highly valued by its owner.

The American continent affords still another proof that the efforts of breeders in various parts of the world have very similar results, although the means adopted and the material employed may vary. In these islands we have had for at least a century, if not as long as pigs existed here, a certain kind of pig which, when young, was in colour of a varying shade of red, and as it reached maturity became of a dark chestnut, and in old age assumed a grizly black tint. Specimens of this or of a similar variety of pig were imported into the States from England and Spain, and afterwards crossed with other kinds of imported pigs, until at last, with careful selections, the so-called Duroc-Jersey breed of pigs was formed. In England we have had exactly the same influences at work, which have resulted in our Tamworth breed. In the main points there is said to be a marked similarity, even in the occasional black spots on the skin and hair, between the red Duroc-Jersey and the red Tamworth, the chief differences being the size and carriage of the ear and the thickness or bulk of the carcase. These variations appear· to have partially come into being from the variations in the standards set up by the breeders in the two countries, but more from the materials at hand and which were used in building up the two breeds. In the Duroc-Jersey foundation a pig of a similar type or character to that which helped to build up the Poland China appears to have been used, since we find the bent or broken ear a peculiarity and an appreciated point in both breeds. Again, the formation of the body of the pigs of both breeds is somewhat similar, and that massiveness, which almost approached coarseness in the Poland Chinas of some fifteen years since, is found in both varieties. Amongst the breeders of the Duroc-Jerseys and the Tamworths we find the same attempts being made to lighten the colour, whilst the admirers of the latter breed are following in the steps of the Americans to widen the backs and add generally to the substance of their favourites. This they have succeeded in doing by the employment of material slightly different in form and character to that utilised by our American cousins; the result has been a change, more marked perhaps in the head than in any other portion of the body; this has been altogether considerably shortened, the nose is less straight, and the jowls are heavier and more full, whilst the small sharp prick ear, indicative of keen hearing, if not of restlessness, has given place to an ear much larger and in-clined to be pendent or drooping forward.

We have heard it asserted that the red pig of the United States *alias* the Duroc-Jersey, is gradually becoming a red Poland China, and that the English red pig will, if the present or just recent steps are continued, gravitate towards the form of a second English variety. We do not quite agree with this contention, since if the Herd Book and its regis-tration are of any real advantage in keeping pure the various breeds of pigs, we look for the Tamworth returning very much to the old form of body, but chiefly retaining the present auburn, or delicate tint of colour of hair and skin.

One of the chief difficulties attending the breeding of cross breds or the altering of the type and character of an old-estab-lished breed does not arise until the home-bred or similarly-bred sires are entirely used; so long as pure-bred sires are used on the cross or improved stock, there is a comparative certainty about the form and character of the produce, but when sires of this newly-formed or amended breed are used on to dams similarly bred, a want of uniformity will generally be notice-able unless in some instances where one particular line of

blood, and that a prepotent one, has been much more extensively used in the manufacture of the new or cross breeds; in
this case the chances are strongly in favour of the produce
partaking and continuing to partake more largely each generation of that particular breed which had furnished the chief or
most important part of the raw material. The reasons for
this appear to be on the surface, so we will not further
pursue this portion of the subject.

So many persons appear to imagine that in the successful
breeding of cross-breds it is only necessary to use parents
possessing a certain proportion of the various ingredients or
characteristic qualities which the animal desired to be produced should possess; or in other words, that the same system
is possible as that pursued by any one in making a compound
article who weighs off a few ounces of one material and a few
ounces of a second, then proceeds to mix well or join the two
together, and the product desired is at hand. Not so, and
particularly, with two so-called pure-bred animals whose particular qualities are, in-bred, but even with cross-breds it is
possible to find a sire and a dam which together appear to
possess exactly those points which we may wish to combine
in one animal, and yet the young produced will not only be
dissimilar to either of the parents, be wanting in several of
the good points, but be possessed of the majority of the weak
points of both sire and dam. Very similar results are sometimes observable in crossing two distinct types or families of
the same breed of stock. Instances of this have not been
infrequently remarked by breeders of Shorthorn cattle who
have used a Booth bull on to a Bates cow, or *vice versa*.
Frequently the produce has been an ungainly animal possessing neither the substance and wealth of flesh of the sire
nor the style and carriage of the dam, or *vice versa*. This is
said to be mainly due to a warring of the two different types
or the qualities possessed by them for mastery over the form
and quality of the produce, with the result that those points
which pertained to the original unimproved Shorthorn and

which were common to the foundation animals of both the
Bates and the Booth types, gained the mastery, owing to
their being natural or more strongly developed than any
separate one of those cultivated qualities which have so
improved the Shorthorn since the earliest breeders took it in
hand. But this angularity and roughness or other failing in
the first cross from the Booth sire out of the Bates dam very
quickly tones down, and is soon lost if a Booth sire is subse-
quently used for two or three generations, whilst the beneficial
effects of a cross are apparent in the increased substance and
stamina of the produce. This leads us up to the point at
which we wished to arrive in connection with the cross-
breeding of pigs, viz., that it is advisable to select for the
purpose two breeds or varieties which have many qualities in
common, or which are supposed to have originated within a
comparatively short period from a somewhat similar source.
The results from the crossing of Tamworth and Berkshire pigs
afford a strong proof of the advisability of following this
system. Although the present style of Berkshire and Tamworth pigs does not possess many points in common, yet it is
an admitted fact that both breeds have possessed in common
ancestors of a similar type. A striking proof is afforded of
this in that the produce of Berkshire boars and Tamworth
sows, if bred together, produce pigs very similar in form and
character to animals of the first cross, *i.e.*, their parents.
The blood blends so much better than does that of two
breeds having nothing in common, or not having comparatively recently originated from somewhat similar sources.

If the experimenter were to breed a Yorkshire boar to a
Berkshire sow the probability is that every one of the litter
of pigs would be white in colour and also partake more of the
form and character of the sire than of the dam. This is said
to be due to the greater prepotency of the Yorkshire pig,
owing to its forbears having been bred of the same colour
for a longer period than the modern Berkshire. Again,
should boars and sows of this Yorkshire and Berkshire cross

be mated for a few generations the resulting produce would become more and more like the Yorkshire, both in colour, form and character. We have seen several instances of this; one large herd belonging to a friend was so bred for a great number of years with the best possible results. This was more particularly noticeable, since the practice had been to fatten the pigs bred. The foundation of this herd was laid some twenty-five years ago by purchase of Large Yorkshires from the author's herd, the produce being reared under the then prevalent system of keeping pigs as stores until they were about a year old, before they were put up to fatten. It was found that the fat pigs became too large and heavy for the local trade, which was the more profitable one, since the breeding of the pigs, and the careful way in which they were fatted, had created a keen demand for the pork which was produced. In order to reduce the size somewhat and still to retain, as far as possible, the colour and quick growth, a pure-bred Berkshire sow was purchased and mated with a Large Yorkshire boar; the half bred sow pigs were saved, and these in turn were put to a Yorkshire, until the proportion of the blood of the White pig would be about seven-eighths of the whole. An attempt was made to use one of the boars so bred on sows of similar breeding, but it did not prove a success, as the produce were not nearly as uniform in size and character—although the colour was invariably white —as were the pigs by a Large Yorkshire sire out of the cross-bred sows. During the last few years a Middle White boar has been used on these sows which had a slight cross of Berkshire, with the result that the pigs are quick growers when young, and then when put up to fatten at about eight months old, very quickly feed to the desired weight and furnish pork of the very best quality. One of the frequent causes of failure in the attempt to breed cross-bred pigs of a uniform type, is the too common practice of purchasing a stock boar first from one herd and then from another, the second often being selected because a slight apparent saving is

being made in the first cost.   This changing about is sure
to result in disappointment—providing of course the herd
from which the original boar is purchased is a really good
and an old-established one which has been managed in a
consistent manner—the pigs will lack uniformity both in size,
style and feeding properties; in fact they will bear more resemblance to mongrels than profitable and comely cross-
breds.

There is little doubt that, given ordinary intelligence,
attention and a persistent line of action and breeding, cross-
bred pigs may in most instances be bred to equal the produce
of many of the so-called pedigree pigs, and far superior in
stamina and general usefulness to a considerable proportion
of those pigs which have a Herd Book number.

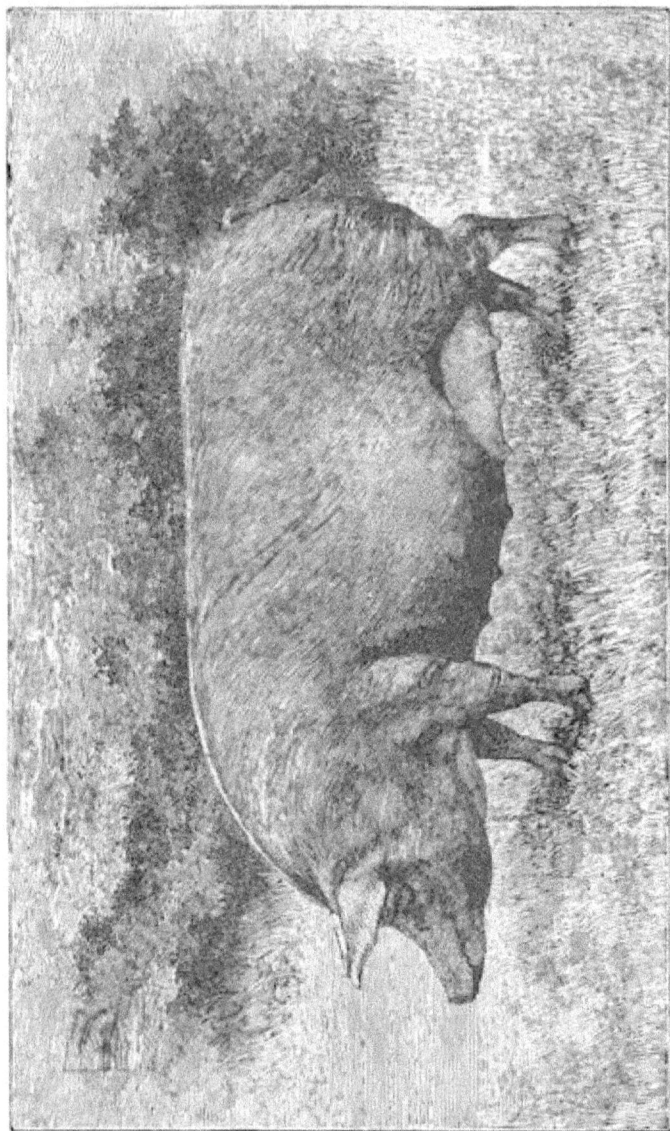

TAMWORTH. SOW, MIDDLETON MAYFLY.

Winner of Prizes at Royal and County Shows.    Bred by and the property of Mr. Egbert de Hamel.

# CHAPTER X.

## HOUSING PIGS.

This, like so many other questions in connection with pig-keeping, is one on which divergent views are held by practical men. This has been more prominently brought out since the abortive attempt was made to stamp out swine fever. A certain few persons hold most strongly to the opinion that swine fever breaks out spontaneously in those wretchedly dirty, ill-kept sties which are far too common in most parts of the country, whilst a very large majority of those who have given time and thought to the subject join issue, and in proof of their contention have not the slightest difficulty in pointing to numbers of the most disgustingly kept sties, which are simply cesspools with a portion covered in, so that only a varying quantity of the rainfall lights on poor piggy's sleeping apartments. In these wretched shanties pigs are found not only to exist, but to thrive, keep healthy and get fat, in direct opposition to all sanitary laws.

We do not for one moment suggest that it is good policy to thus neglect the housing of one's pigs, but simply mention the fact to show how trivial need be the outlay in building a sty in which successful pig-breeding and feeding can be carried on. We have in our time seen some of the most elaborately designed sties, in which it was simply impossible to keep and rear young pigs. In one instance present to our mind the architect appeared to have carefully considered the requirements of every kind of stock save one; the food-preparing machinery, the stables, cow byres, calf pens and bullock boxes were all most conveniently arranged before the slightest thought was given to the housing of the good herd of pigs, which was kept mainly for the purpose of supplying the hall and home farm with pork. Fortunately or the reverse there appeared to the architect no necessity for a re-modelling of the premises, since there was an un-occupied portion on the north side of the buildings, which latter happened to be so high that the sun never by any chance shone on or into the sties, which were made in this unsuitable position; then, to mend matters, the partitions of the sties were of sheet iron, and the floors laid with asphalte. Little wonder that under such conditions the young pigs invariably suffered from diarrhœa and rheumatism, and persistently refused to thrive and grow. Crampy, stunted, ill-thriving pigs were the order of the day, until at last the pigman in despair turned his down pigging sows into a large yard with an open shed, which was usually used for the colts in spring time. Here the sows farrowed, and although it appeared at first to be somewhat cold for the newly-born pigs, yet after a week or two the youngsters grew and throve in a manner which surprised the farm steward, who had not hesitated to lay a considerable portion of the blame for the previous want of success on to the boar and the sows, which were not of the breed to which he had been accustomed in early life. However, so convinced did the owner become of the unsuitability of the grand sties that they were forthwith closed and others built somewhat apart from the farm premises. This unsuitability of expensively built piggeries is by no means uncommon, especially when the architect or other person who has drawn up the plans has had no experi-ence of pig-keeping, and has not deigned to seek advice of those persons who have had this experience. Again, so many times uniformity of appearance in the general buildings has been considered, instead of such a simple but yet important

matter as aspect. The sties for the sows and young pigs
have been made to face the east, the north, north-east or
north-west, or exactly those points of the compass from which,
the cold winds are frequent in the winter and spring, which
with the want of sunshine, are fatal to the well-being of
young pigs. Of course for fat pigs in the summer time these
places would be suitable, or less unsuitable than for little
pigs, still one may just as well so arrange when building
piggeries that each sty shall be suitable for any kind or age
of pig at all times of the year. This is quite possible, and
can be done at no great outlay. So far as our experience
enables us to form an opinion, the best aspects for piggeries
are south or south-west.

With regard to this, and to the slight necessity for
expensively built piggeries, we had a somewhat unique
experience when awarding the prizes offered by a local
authority for the best pigs of certain ages, the property of
labourers or artisans in the district. There were some forty
entries, and as many of the owners had entered only one pig
we necessarily had to view twenty to thirty pig sties, since
the pigs were not collected together, as is usual at shows,
but each entry was inspected in the sty it usually occupied.
These were of all sorts and sizes, some of them had evidently
cost very little in their construction, being merely a few old
railway sleepers and some odd boards nailed on them, and
in some cases the roof was simply bushes or hedge-side
trimmings thatched ; but it was noticeable that in nearly
every case where the pigs were thriving and apparently
profitable, the sties faced the south. We were surprised to
find several litters of young pigs as blooming and healthy as
possible, although the inspection was made early in the
month of March. These trips of youngsters afforded a great
contrast to the stunted, unhealthy little pigs which we have
frequently seen in some of the most expensively built
piggeries at some home farms, and even on the premises
of amateur farmers who simply followed the pursuit of

8

farming as a pleasure or hobby, rather than for profit, so that the question of expense did not arise or interfere with the feeding arrangement or management.

It is most probable that one of the principal reasons for the pigs in these amateur-built sties thriving so much better than those more expensively fed in grand piggeries, is the thorough ventilation which is unintentionally afforded, owing to the impossibility of making the sties comparatively air-tight, as are those sties brick built, slated, and with carefully made doors and windows. It is true that many persons imagine that it is necessary to have an outside court in which to feed the pigs, and to which the latter can at certain times adjourn; an opening into the sleeping compartment is made, and most frequently this is without any door or other means of excluding the draughts, which are never absent when there is a disturbance of the atmosphere. Sties thus built have plenty of ventilation in the cool weather, at the time when it is perhaps less needful, but they are also so cold and draughty in severe weather that a considerable amount of the food consumed by the occupants of them is required to keep up the animal heat. Another frequent failing of these piggeries is the want of height in the side walls; the sties so built quickly become insufferably hot in summer, and in winter time are either very cold or unhealthily close when the openings are closed. The reason for so building these piggeries, which are far from healthy, is doubtless, an attempt to save expense in building, but if the extra cost of one, two, or even three feet of side wall is placed against the loss of thrift of every pig, or lot of pigs, kept in such sties, the loss far exceeds the amount which should have been expended in increasing the height of the building, and rendering it far more healthy and less subject to variations in temperature and to draughts.

Some of the very best piggeries are amongst those converted out of old-fashioned barns; these are far sweeter, and altogether better for sows suckling and for young pigs. The temperature is equable and the freedom from currents of air particularly noticeable. The majority of these barns are thatched with straw or reeds, so that they are warm in winter and cool in summer. Where the roofing is of tiles it is in the winter a good plan to fill up the roof of the barn with straw; this does not interfere with the ventilation, but renders the place much warmer and more free from draughts.

Some of the so-called model piggeries, erected at a large cost in various parts of the world, have a double row of sties with a walk down the centre, the partitions of the sties being some three to four feet high, and doors opening outside as well as into the passage. When the side walls are sufficiently high to allow of a plentiful supply of fresh air without draughts, this kind of piggery answers well for pigs of six months and over; but if the buildings run from east to west the sties on the south side only will be suitable for sows and young pigs; those sties facing the north will prove but very moderate places for young growing pigs in the winter or early spring. Others of these double piggeries are planned to run from north to south; here, again, we have one-half of the sties facing the cold east, and the other the variable west, and neither so placed that the occupants of the sties receive any large amount of the benefits derivable from sunshine, which is indispensable in the successful raising of young pigs.

Perhaps we could at Holywell Manor furnish a greater variety of styles of pigsty than any other breeder of pigs. We have over sixty sties of all sizes and forms. An old barn is converted into eight sties, a lofty shed also furnishes eight, whilst a second shed is fitted up with eight sties in which the sows farrow during the colder weather. As the autumn approaches the roof is crammed with wheat straw, which renders the place very warm and yet airy; the front of the shed is boarded up, in which there are two double doors and ventilators, so that the temperature can be fairly well regulated, considering that we have no means for artificially heating in severe weather. In the back part of the shed are two doors which are set open in summer, so that a thorough draught is

obtained without the occupants of the two sets of four sties
being affected injuriously.  Then for the young boars and the
growing pigs we have a row of eight sties, 10 feet by 11
feet ; these face the south ; at right angles with this row is
another of seven sties, 9ft. by 11ft., facing the west.  These
sties are all 7ft. on the side walls, have double doors, and
ventilation fore and aft ; so that in the very coldest weather
these can be closed and the sties rendered sufficiently warm,
whilst in summer a current of air is obtained above the pigs.
The partitions are only about 4ft. high ; in the winter time
some of these are extended to the roof by nailing up old bags,
but in hot weather these temporary divisions are removed,
and thus the whole of the sties are rendered cool and sweet.
Some buildings close in the west side, so that a square some
30 yds. by 25 is formed, in which each of the various lots of
pigs are allowed to take exercise on most days for a short
time, and are thus kept healthy and hardy.  It is asserted,
and not without a good deal of truth, that more pigs are
injured by want of exercise than from having too much
freedom.  In winter and early spring young pigs frequently
lie undisturbed for fifteen hours, curled up in a heap, and often,
if much litter is given to them, pretty well embedded in it.
Over-feeding and the general want of exercise frequently
result in the little pigs suffering from what is commonly
termed cramp, but which we are inclined to think is merely
rheumatic gout.  One of the most healthy of our sties is one
which is simply a loose box, being the end of an open shed,
facing the west, and which was evidently intended as a place
in which young calves might be kept or a heifer might be
placed to calve ; the front is boarded, in which is a door, and
the wooden partition dividing the place from the open shed is
as high as the side walls, so that the roof of the south end
opening into the shed is quite open.  It would thus appear to
be a somewhat cold place, whereas it is, in fact, so healthy a
one that it is called the hospital, and any little pigs not thriving well—and there are certain to be some of this character

where sixty or seventy sows are kept—are removed to this hospital for a time. This would tend to furnish proof that it is not warmth alone which little pigs need so much as plenty of fresh air with freedom from draughts.

It is considered by many persons who buy store pigs for fattening that it matters but little as to the kind of sty or of the aspect; they will argue that the one thing needful is a plentiful supply of good food. There is little doubt that in this country the housing of the fatting pigs is not of so much importance as in colder climates, still the exposure of the fatting pigs to extremes of heat in summer and cold in winter is certain to result in a waste of food. In the British Isles the winters are not usually so severe for any length of time as in the States, where experiments have proved that in the very cold weather fatting pigs have for days at a time made no gain in weight, the whole of the food having been required to keep up the heat of the pig's bodies; whilst in other experiments the increase in weight from the consumption of a given weighty meal has been regulated by the degree of cold to which the pigs were subjected. It is thus clearly proved that in cold climates the necessity exists for constructing the sties for fatting pigs in such a manner that the temperature of the sties cannot fall below a certain point. There is little doubt that even in this country the fatting of pigs would be more profitable if the buildings in which the pigs were housed were so arranged that the occupants were not subject to extreme or even slight changes of temperature. It is very probable that in the near future the fatting of pigs in the summer months will be more general. In the corn-growing districts the system has been to confine the fatting of pigs mainly to the winter months, or when the inferior corn grown on the farm was available, whilst the summer fatting of pigs was far more general in the dairying districts, when the pigs were required to convert the dairy offal into pork. In the latter case the pigsties may be arranged with little thought as to the aspect, since in the very hot weather the fatting pigs will actually thrive better without the warmth of the sun's rays than when the sties face the south ; whereas in those cases where winter fatting is followed it is imperative that the sties should be so arranged that the temperature can be regulated, in order that an undue portion of the food eaten by the pigs is not required to keep up the natural heat of the body.

It is quite possible to erect piggeries having the best of ventilation, and all those other necessary requirements, and yet, if the site selected is lower than the surrounding ground, or if the land abounds with surface springs, the young and growing pigs will not thrive. They will be liable to rheumatism, colds, &c., or affected in a similar way to those hounds which are subjected to the same influences, *i.e.*, an affection which is termed kennel lameness. The little pigs will become crampy and unthrifty generally, and even sometimes older pigs, which have been brought from sties built on land thoroughly drained, either naturally or artificially, will fail to thrive. Although the pig has a weakness for a mud bath in hot weather, yet this is only beneficial—if at all—when enjoyed for a limited period. A dry lair is necessary, particularly for young pigs in the autumn and winter months. To carry off any superfluous moisture, surface drains are infinitely superior to gratings and pipe drains, as the latter are a continual source of worry through the short litter getting into and stopping them up.

# CHAPTER XI.

## EXPERIMENTAL PIG-FEEDING.

In this country, which undoubtedly possesses the most valuable breeds of pigs in the world, we do not appear to have taken anything approaching the trouble which several of our foreign competitors have done to discover the best and cheapest kinds of food on which to fatten pigs. It is true that, some years since, Sir John B. Lawes—to whom all stock-feeders are immensely indebted for the enormous amount of time, care and money expended on stock-feeding experiments carried on at Rothamstead—turned his attention to the discovery by actual results of the feeding values of different kinds of grain for the fatting of pigs. This series of experiments were most valuable as far as they went, but in Germany, Denmark, the United States, and other countries, there are Agricultural Colleges, Schools, or other Governmental establishments in which experiments in the feeding of the different kinds of animals are almost continually being carried out and the results published at the public expense. In the British Isles we appear to have rested content with the superiority of our pigs and of our pork products, a superiority which may be due, to a certain extent, to causes which may in the near future have a similar effect on the produce from the pigs of our rivals, when we shall perhaps awaken to find that the markets for the finest bacon and hams have, like those for high-class dairy produce, been captured by the enterprising foreigner. There is little doubt that the breeders and feeders of pigs, as well as the bacon-curers in Denmark have made enormous strides during the last ten years towards the production of hams and bacon which compare favourably with that Irish bacon for which its manufacturers justly claimed the credit of being the finest in the world.

There is another point which we in the British Isles do not generally appear to appreciate at its full value ; it is that the pig which is to furnish the style, form and substance of the side of bacon now most in demand must not only be bred on different lines, but that the system of feeding which was in vogue some twenty-five years since is totally unsuited for the production of that carcase of pork which, when converted into bacon, will command the highest market price. In the good old times the plan was to allow the pig to be a store, or in other words a more or less starved, animal for at least twelve months, after which it was put up to fatten for a period of four to six months, when it consumed an enormous quantity of food, much of which was required to enable the poor pig to regain that flesh which it ought never to have lost, and which might have been retained at a tithe of the cost of renewal had the animal been even humanely, not to mention judiciously, fed. The resultant fat pig was an immense animal, strong in bone, coarse in hide, and several inches thick in fat on the back and ribs. This mass of fat might have been necessary in the consumption of this large, heavily-salted bacon, since it would be impossible in the present day to secure anyone to consume the fearfully hard and extremely salt lean flesh of the pig of that period. The increased amount of wages earned by our artisans, and the great change which took place in the style of their living, very quickly had its effect on the kind and class of meat consumed by the so-called lower classes. Fresh meat and frequent small joints took the place of few large joints of heavily-salted meats. An idea has been expressed that, had it not been for this change in the habits and requirements of our labourers, we should still be employed in literally wasting a

large proportion of the food fed to our fat pigs, as our fore-
fathers did when they kept their pigs twice as long as they
ought to be kept before being slaughtered. This may be so,
although we doubt it; the marvellous improvements and
increase in our stock papers, and the publications of the
proceedings at the foreign Governmental Agricultural Schools
and Colleges, added to the increase—totally inadequate though
it be—in the number of agriculturists who read, have all had
a share in causing greater attention to be paid to the more
general study of judicious and economical feeding of stock,
particularly of pigs, which in the good old times were too
frequently looked upon as necessary nuisances.

With regard to the question of the amount of food required
at the different stages of fatness of the pig to produce a given
increase of weight, in years gone by most farmers would have
asserted that a certain weight of food would, under ordinary
conditions, give an equal increase of weight if fed to pigs
weighing 400lbs. as if fed to one of 150lbs.; whilst some
farmers would go even further, and declare that a greater
increase was obtained by feeding a certain weight of meal
to a large pig well-nigh fat than to a comparatively small
one not in high condition. If asked to state the grounds
on which they had come to this conclusion, their reply
would be that their experience had proved such to be the
case. No attempts would have been made either by weighing the food consumed or the animals at the various stages
of fatness; the eye alone was trusted to, and in this, as in many
other instances, the eye had deceived the owner.

Carefully-conducted experiments have clearly proved that as
the weight of the pig increases so does the amount of food
required to produce a given increase become larger. At the
Danish State Agricultural Experiment Station experiments
were carried out which proved that with pigs weighing from
75 to 115lbs., it required 4·37lbs. of food to produce 1lb. of
increased weight; from 115 to 155lbs., 4·67lbs. of food; from
155 to 195lbs., 4·99lbs. of food; from 195 to 235lbs., 5·43lbs.

of food, and from 235 to 275lbs., 6·24lbs. of food were needed, or nearly half as much again corn or its equivalent was required to add 1lb. to the weight of a pig weighing 235 to 275lbs. than when the same pig weighed but 75 to 115lbs., young and in its early stage of fatting. One of the chief causes for this extraordinarily increased quantity of food required for the addition of 1lb. of weight to the older pig doubtless is that the young pig, which is in a growing state, is able to utilise more or a greater proportion of those food constituents which go to build up the frame, whilst the fully-grown pig can increase its weight only or mainly by the utilisation of those other food constituents needful for the production of meat, or, more strictly speaking, fat.    This alone would render the fatting of young and immature pigs more profitable than fully-grown pigs in a fair state of fatness.

The figures given in the report of the experiments would, of course, require some modification, according as the pigs were of a small and early maturing kind, or of a large variety which would grow to a larger size and come later to maturity.    This being the case, it would appear to furnish another argument in favour of fatting animals of the larger breeds, providing only that they possessed a fair amount of that which is generally termed quality, but which should be fineness of bone, skin and hair, and ability to turn the food consumed into the largest quantity of the most saleable kind of meat.

It would thus appear that the present demand on the part of the consumer for small joints of meat off quick-growing animals of the larger type is really a blessing in disguise to the feeders of stock, who at first were inclined to resent the demand of the public for meat of this character.    Not only so, but the breeders of our domesticated animals must share in the benefits which will arise, since a larger supply of animals must be produced in this country.    This will be the more readily and profitably carried on now that there is no risk of the importation of foreign diseases.    There is but little doubt that the actual and anticipated losses resulting from imported diseases have contributed very materially to that depression in agriculture, which has wrought such havoc amongst landlords and farmers, but which we really believe is becoming less acute than it was some one or two years since.

In an attempt to determine the amount of food required in summer and in winter to produce a given increase, 100 winter and 99 summer experiments were carried out. As each experiment included from 25 to 30 animals, the large number of over 5,000 pigs were utilised. This should be sufficient to render the results of the experiment convincing. It appears that the animals ate very little more in winter than in summer, but that 4·4 lbs. or nearly ½ lb. more food was required to produce 1 lb. increased weight in each pig in winter than in summer. This may not at first sight appear to be a very strong point in favour of summer as compared with winter feeding ; but if we take 5 lbs. of corn, or its equivalent, to be required in summer to produce a gain of 1 lb. of weight, and if 5½ lbs. are needed to produce the same result in winter, we save at least 8 per cent. of food by fatting our pigs in summer rather than in the winter. Not only so, but the average price of pork is said to be higher during the summer months, and less straw or other material for bedding is required in summer than in winter, and this is an important item in many parts in the north of England.

These experiments afford strong evidence in favour of the contention long held by some of us that our general system of winter pig fatting was not now so profitable as the fatting of pigs in summer could be made. The more general plan of winter fatting of pigs had, in times gone by, reason on its side, since before the present system of mild curing bacon was adopted, the chief demand for fat pigs in the summer time arose from the butchers, who retailed the meat as fresh pork. Again, the system of keeping pigs during the whole of the summer, running about as stores and the consumers of uncon- sidered trifles about the farm, and then to be shut up in the

autumn to convert the inferior or tail corn into meat, had in
times past much to commend it. But we have changed all
this ; not only so, but the major portion of the pork now produced in this country is made from the consumption by pigs
of foreign grain, whilst it is probable that in the near future a
still greater proportion of the pig food will be imported, as
under ordinary conditions the purchase by the farmer, and the
consumption of foreign corn by home-bred stock, is considered
by many persons a less expensive and more thorough way of
restoring or increasing the fertility of our arable and pasture
land, than by the application of the so-called light artificial
manures.

The effects of the two systems of feeding, which are sometimes called light and heavy, *i.e.*, the giving of a fair quantity
of food, or the cramming of the fat pigs, were incidentally
observed in a large number of experiments ; but no marked
difference was noticed between the amount of food required
per lb. of gain on light and heavy feeding. From this it
would appear as though the slight amount of food saved on
the heavy feeding, by the pigs becoming fat in a rather
shorter time, was equalised by the pigs utilising a somewhat
smaller proportion of the feeding properties of the food
consumed.

An attempt was also made to determine which of the two
sexes, male or female—after being operated upon—proved the
quicker and the more profitable to fatten. It was found that
the return for food consumed was about equal between the
boar pig castrated when young and the sow spayed at an
early age ; but an advantage rested with the females, inasmuch as their carcases proved as a whole to be more valuable.
Thus 56 per cent. of the sow pigs furnished carcases graded
as first class, whilst only 44 per cent. of the carcases of the
barrows, or castrated boar pigs, were placed in this category,
whilst 85 per cent. of the sows and .77 per cent. of the barrows
were, when killed and dressed, considered to qualify for
positions in the first and second classes. This may not be an

inopportune moment to point out how necessary it is to have the sow pigs spayed; the food consumed when fatting will be less, the quality of the meat better, and the spayed pigs can be marketed at any time, since they are in a condition fit for killing. The losses sustained by bacon curers and others, who wish to salt the pork, from the killing of non-spayed sow pigs, is very considerable, so that all persons who handle these neglected pigs suffer a loss.

Five series of experiments with 115 pigs were made to discover the feeding value of barley and maize, alone and mixed.

Lot 1 were fed on barley alone.

Lot 2 were fed first on maize, and then when they weighed some 120 lbs. each, barley was substituted for the maize.

Lot 3 were also at first fed on maize, and this was continued until the pigs weighed 140 lbs., when barley took its place.

Lot 4 were similarly fed until the pigs reached 160 lbs., when they, too, were fed on barley.

Lot 5 were fed on maize throughout the experiment.

Those pigs fed on maize alone made somewhat greater gains than did those pigs fed on barley alone, but the quality of the pork was not so good, as proved by 92 per cent. of the carcases of the barley fed pigs being placed in the first two classes, whilst only 62 per cent. of the carcases from the maize fed pigs were thus placed, and still worse, 14 per cent. of the latter carcases came within class four, which comprises poor carcases sold at a much lower price. The softness of the pork increased in proportion to the quantity of maize fed. Thus if we take one as perfection, the softness of the pork from the different lots was as follows :—Lot I., 1·4 ; Lot II., 1·6; Lot III., 2·0 ; Lot IV., 2·3 ; and Lot V., 2·7.

# CHAPTER XII.

## PIG CENSUS AND VALUE.

IN endeavouring to form an estimate of the variations in the value of the pig stock in the United Kingdom on June 4 in the last six years, we are met with several difficulties which render the task a somewhat difficult one, and at the best the estimates may be only approximately correct. In the first place, the Agricultural Returns simply give the gross number of pigs on occupations exceeding half an acre, and until within the last three or four years, after the suggestion was made to the Board of Agriculture by the author, no attempt was made to give a summary of the numbers of breeding sows and other pigs at a stated time, whilst even now no indication is given of the numbers of pigs above or under six months old on each fourth of June. An instance of the way the value of the pig population may be affected by the want of division, might easily be furnished by simply pointing to the disproportionate number of sows and of older pigs compared with the number of smaller and younger pigs at those times when pigs are increasing or decreasing in value. Thus in June, 1891, when pigs were exceedingly cheap, the size and weight of the pigs then in the country would be very considerably greater than in June, 1892, as at the first period numerous owners had then for some time been holding on their fat and older pigs in the hope of selling them at a higher price, and also because of the slight demand for fat pigs and pork which had for some months been experienced. Exactly the opposite conditions were in force about June, 1892, most of the owners of fat pigs and a very large proportion of those who usually keep sows for breeding purposes had cleared all out in the early spring, and the demand for pork was so keen in the early summer that those persons who are in the habit of feeding their pigs to great weights, had been so tempted by the high prices offered for porkets and small pigs that they sold out and had on hand smaller and less valuable pigs.

Again, this wonderful increase in the value of pork and pigs, and the equally large reduction in the price of feeding stuffs, led to largely increased numbers of sow pigs being left unspayed, and then kept for breeding purposes.

The proportion of pigs in various parts of the country also varies very materially; thus when the artisans and colliers in the north are in full work at high wages, numbers of them will take to pig breeding, and in those districts the average weight of the pigs slaughtered is very considerable—at least 50 and mayhap 75 per cent. greater than the average weight of the pigs killed in the southern counties, where more pigs are kept when prospects are not bright than in times of plenty, and as these pigs are held by occupiers of less than half an acre, no account is taken of them in the pig census taken on June 4 each year. This is partially due to the lower values of pig feeding stuffs, and to the greater demand for pork, because when wages are lower the farm labourers and artisans consume a greater quantity of pork and less of beef and mutton.

The change in fashion as to the size of the joints now found on the tables, especially of the middle classes, and the enormously increased demand for, and value of small hams, of, say, 10 to 12 lbs. weight, compared with the old-fashioned 30 lbs. hams, and the enormously increased consumption of mild cured bacon have tended very materially to the slaughtering of pigs at a much younger age than was usual only a few years since.

These and many other changes which have taken place, and the omission in the returns of the estimated ages of the pig population of this country may, in the opinion of many persons, rob the figures which might be given as to the variation in the value of the pigs in the British Isles at various periods, of a great portion of their value, and reduce the estimates almost to the level of guesses. Nor does it appear that any attempt to show by figures how much our pig-keepers may have lost or gained each year by the variation in the value of their pigs is likely to prove of much benefit. Every one who has given the slightest attention to this pig question is well aware of the considerable fluctuations in the value of pigs, and further that these changes take place well nigh periodically. Every four or five years we find pigs are said to be gold or copper, and just as regularly do we find thousands of people hastening to get rid of their pigs at any price, and then in a year or so equally as anxious to become the possessors of breeding sows at any cost. The marvellous prolificacy of pigs is of course one of the chief causes of this great variation in their value, but we are inclined to think that it is also assisted by the very small number of bacon curing establishments in England to deal with the supplies of fat pigs. It appears strange that in this country, where the con-sumption of bacon and hams is so enormous, that bacon factories are almost unknown, except in the South-Western counties. It may be that the imports of fat pigs from Ireland added to the home-fed pigs, furnished that regular and con-tinuous supply of fat pigs which is necessary to carry on profitably the curing of bacon on a large scale ; but of late years thousands of pigs have been sent from the Midland and Eastern counties into the counties of Wilts and Somerset. Not only so, but the general quality, form and size of the Midland counties' pigs are said to be exactly such as are required in the manufacture of the most valuable bacon and hams. If this be the case it appears strange that the bacon curing industry has not been successfully followed in the

Eastern Midlands, since labour is nearly as cheap as in Wiltshire or Somersetshire, whilst the best markets are quite as easily reached. Indeed, an advantage in this respect rests with several counties where pigs are very largely kept. Take, for instance, the three Eastern counties of Essex, Norfolk, and Suffolk, and we find that in these counties about one-fourth of the pig population is kept. Surely it is a mistake that some steps are not taken by those interested in the securing of one or more certain markets close at home, where it is possible to dispose, at fixed prices, of these fat pigs. It is true that two attempts have been made to establish factories. One of these was started in Suffolk, in which county, unfortunately, the common pig of the country is by no means suited for conversion into the highest-priced bacon ; consequently the produce of this bacon factory had to come into competition with bacon produced abroad, and which had been manufactured from pigs that had cost little to breed and fatten, and that, therefore, could be sold at a profit to the curer at a much lower price than it was possible to produce fat pigs in an English county. It was also stated that the directorate were not practically acquainted either with the manufacture of the bacon or with the disposal of it. If a factory had all these disabilities to contend with, it can scarcely be a matter of surprise that it should prove to be an unprofitable investment.

The attempt made to establish a bacon factory in Norfolk was an abortive one. There did not appear to exist grounds for any great complaints as to the style and formation of the fat pigs, nor as to the difficulty of finding a supply at reasonable rates. This want of success was thought to be owing to the system of buying the fat pigs not being exactly suited to the conditions existing in the local districts. This trouble was not lessened by the system of selling most of the fat pigs at the auction marts, some of these sales being—we understand—attended by some one connected with the factory, with the natural result that all the dealers were united as one man

9

in rendering it impossible for the agent of the factory to purchase any number of pigs without the assistance of the dealers,
except at a considerable advance on the then current value of
fat pigs.  We might remark in passing that these weekly
auction sales are said to be responsible, to a very great extent,
for the excessively low or high prices of fat pigs.  It appears to
be tolerably certain that in times of over-supply of fat pigs the
auction sales place in the pig-dealers' hands a most powerful
weapon, which they not infrequently use for their own aggran-
disement and to the great loss of the pig-feeder.  This unfortunate state of affairs would be rendered less acute if bacon
factories were more general, and if the system were adopted,
which has been found so satisfactory by Messrs. Harris, of
Calne, of pig-feeders and others sending the fat pigs direct
to the factory where they are slaughtered and weighed, and
the current price forwarded to the owner.  The plan of esti-
mating the value of the fat pigs according to their weight,
substance and form, tends greatly to encourage pig-feeders to
breed and fatten the very best kind of pigs, since any benefit
arising therefrom is at least shared by the owner of the fat pigs.
Still another encouragement is given by Messrs. Harris & Co.
to breeders and feeders of fat pigs to send to them animals
of the best quality, by paying to the owner a bonus of 2s. 6d.
on each pig which comes up to a certain standard of merit.
The way in which Messrs. Harris & Co. at present arrive at
the value of the fat pigs consigned to them is to classify them
as follows :—

" Pigs when dressed and weighing—

| Prime stores. | Thickness of fat in any part of the back. |
|---|---|
| 6 sc. 10 lb. to 9 sc. 10 lb ... ... | 2¼ in. and under |
| Under 10 sc. 10 lb. ... ... ... | Not exceeding 2½ in. |
| Under 11 sc. 10 lb. ... ... ... | Not exceeding 2⅞ in. |
| Under 12 sc. ... ... ... ... | Not exceeding 3 in. |

" Any pigs outside these limits at their value.  Half-truck,
12 pigs ; whole truck, 25.—*Chas. and Thos. Harris and Co ,
Limited.*"

The further conditions are that they shall be prime fat pigs in lots of not less than 10 on rail within 100 miles of Calne. This limit of 100 miles is not intended to preclude pigs from a greater distance being sent to Calne, but simply that the consignor is to pay the extra cost of carriage charged on consignments from a longer distance. It appears that by this arrangement the feeder who supplies the best fat pigs cannot fail to reap a considerable benefit, quite sufficient to pay for the extra cost, care and trouble, which are required to produce the best description of fat pigs. We are strongly of opinion that the steps taken by Messrs. Harris & Co., some fifteen to twenty years since to bring home to pig-breeders the fact that the then fashionable pig was a most unprofitable one alike to producer, bacon-curer and consumer, and the system of purchase to which we have referred, have resulted in enormous benefits to pig keepers generally. The little fat puggy pig has, like the great, coarse, overgrown pig, well nigh disappeared, and quick-growing, early-maturing pigs have taken their place.

If we read aright the signs of the times, a similar great change is taking place in the style, feeding and formation of the pigs in the United States. The American pig has ceased to be either an immense creature suited only to be converted into salt pork or lard, or the little animated bladder of lard. It is true that the interests of the owners of those herds of pigs of the breeds which have been fashionable in the States for a number of years are thought to be attacked by those advisers who call attention to the fact that the heavy fat pig must give place to the larger and leaner fat pig of not more than nine months old. Millions of pigs have annually been fattened on Indian corn alone and at very little expense; the resultant carcases of pork were far more easily disposed of in the sixties than now; indeed, the manufacture of lard has from various causes, ceased to be profitable and has actually been declared to be of less value on the market than some of those many forms of United States manufacture which have done duty for lard. Attention was recently called to the fact in the columns of one of the leading stock papers in America.

Not only has the age at which fat pigs have been killed, and the average weight of the pigs killed been reduced, but the pig population of the United States is at time of writing nearly thirteen millions fewer than at the highest point, when the estimated number was some 53,000,000. It is true that at the present time the average duration of life of the pig is shorter than it used to be in the States, but owing to greater care in feeding and the attention paid by breeders of pigs to early maturity, the average weight of the nine-months-old fat pig is increased. Still the withdrawal of eight or ten millions of pork makers within a few years must affect the supply of pork in the States, even after every allowance has been made for the reduction in the quantity of pork-eaters *per capita* in America. On the other hand it may be said that the population in the States is increasing at a rapid rate, and that times are so bad generally that many persons who have given up the eating of pork for the more fashionable flesh of cattle and sheep, will now be compelled to return to their cheaper meat-food, pork.

In this country few persons appear to realise the immensity of the American export trade in pork products. A return recently issued shows that there was a slight falling off in 1896 as compared with the exports of the first seven months in 1895. The actual weights and values are—of bacon 241,812,772 lbs., valued at 17,610,923 dols. ; of hams, 89,744,920 lbs., valued at 8,718,175 dols. ; of pork, 35,171,080 lbs., valued at 1,862,220 dols. ; and of lard, 264,520,015 lbs., of the estimated value of 16,106,581 dols., or a total of 631,248,727 lbs., valued at 44,302,949 dols., or something like £9,000,000 of pork products shipped during seven months. If we take an average export for the remaining five months of the year, we arrive at the enormous sum of about £16,500,000 as the value of the pork products exported from the United States in one year.

Compare this with the whole of the yearly produce from our small total of less than 3,000,000 pigs in Great Britain. It is doubtful if the value of hog products produced in Great Britain and Ireland during the last twelve months has exceeded half the value of the exports from the States. Of this export we have doubtless had the lion's share, since the regulations in force in Germany and France, and the high import duty in both these countries, have helped to make the British Isles by far the best market to which the Americans could ship their surplus pork products.

Estimates have been made at times as to the value of our pigs and of the yearly produce and its value from them ; but these have varied so considerably that the public have hesitated to place any great faith in them. The wonderful change which has of late years taken place in the system of management and of the style of the pig, and its age when it is made fit for slaughter, have all had their effect on the amount of meat produced yearly from a given number of pigs, as well as on the average value per head of the pig population.

The returns for the year 1896 show but little variation in the gross numbers returned in the year previous, but there is one rather ominous change in that there is shown a reduction of some 21,481, or rather over 5 per cent. in the total of breeding sows. This points to a decrease of something like 150,000 pigs at the next census to be taken in June, 1897, by which time we shall in all probability find that the value of pigs and of pork produce has considerably increased.

## CHAPTER XIII.

### DISEASES OF THE PIG.

In the pig, as in some other animals, domestication and its attendant infringement of nature's laws have largely increased the susceptibility to disease, and there is reason to think that the number and variety of his ailments have also been considerably added to in consequence. Notwithstanding this, pig health and pig diseases have received but scant attention from the veterinary profession, nor does the fact excite surprise when it is remembered how reluctantly in this connection the services of the veterinary practitioner are enlisted save when serious losses are threatened by the existence of contagious disease.

In the absence of a systematic treatise on the diseases of swine, stock owners will find in the following pages a concise account of the nature and origin of the more common ailments of the pig, with a description of the measures necessary to their rational treatment and prevention.

#### Swine Plague.

Swine plague is a specific fever which for many years was regarded as a form of anthrax, both in England and on the continent, until 1877, when the writer showed it to be a distinct disease, communicable from one animal to another in one of several ways: (1) through the atmosphere; (2) by means of food; (3) by inoculation. The virus is contained in the excrement, in the saliva, in the discharges from the skin, and in the blood and tissues of the body generally. It has since then been shown to be due to a minute organism possessing considerable vitality, and capable of retaining its virulent properties for a long period. The discharges from the bowels are the most fruitful source of infection, hence it is that markets, sties, yards and roads become contaminated and rendered so liable to extend the disease to healthy stock herded on, or passing over, such infected ground.

Young animals show a greater susceptibility to the disease than older ones, and when closely herded together quickly become affected. Animals of all ages, however, and in all conditions are liable to infection.

*Symptoms.*—The symptoms of swine plague are usually declared in from four to eight days after infection. In some cases a longer period elapses, but this is the exception. The period of incubation appears from our own experiments to be governed in some measure by the quantity of virus received, and this view has been strengthened by competent observers since Dr. Klein affirmed the contrary. Where many pigs are herded together, and the food becomes largely contaminated with the poison, the incubation of the disease is then curtailed, and its course becomes rapidly fatal.

The duration of swine plague varies with its severity. In some instances it kills in two or three days, in others it extends over weeks. Moreover, there is reason to believe that under some circumstances the action of the poison is very feeble, and animals, although affected with the disease, exhibit no perceptible indications of it, and consequently escape detection.

The first notable sign of illness is a rise of temperature. Then follows more or less reddening of the skin beneath the belly, at the root of the ears or on the under side of the arms and thighs. The rash is sometimes in patches, at others spread over the entire body, and it may or may not be attended with an eruption of small vesicles or blisters and the subsequent formation of scabs. At this time the pig is dull, and the ears and tail droop; food is refused or taken sparingly, and there is a general expression of sickness.

The patient buries itself in the litter in search of warmth, and is with difficulty made to move. If several are affected at the same time they huddle together. When standing, the back is arched, the belly tucked up and the flanks are hollow. The eyes discharge a watery fluid which later on becomes thicker and causes dirt to adhere about the lids. A similar discharge flows from the nose, and implication of the lungs is shown by a frequent cough. As the disease progresses the movements are feeble and unsteady. In some instances the animal becomes stupid, giddy, and even delirious, when, without apparent reason, it squeals and starts as if frightened. Paralysis of the hind quarters is frequently seen, and this may be attended with convulsions.

In the early period of the attack the bowels are constipated, but later on diarrhœa sets in, and wasting and fatal exhaustion follows.

### Apthous Fever ("Foot and Mouth Disease").

Foot and mouth disease is a specific contagious fever. Its chief characteristic is an eruption of vesicles or blisters on the snout, lips, and feet. Rarely they also appear on the udder in sows when suckling. All our farm animals are more or less susceptible of the disease, and we have on several occasions known it to occur in the human subject as the result of drinking specifically infected milk. Its presence on a farm, therefore, calls for the greatest vigilance and care.

*Cause.*—The contagion in this disease, as in anthrax, tuberculosis, and some other spreading affections, is believed to consist of a minute organism, but at present its identity cannot be said to have been clearly made out. The spread of the malady is favoured by anything to which the virus may cling, or with which it may be mixed. Rats and poultry, frequenting pig styes frequently convey it to byres and pastures. Manure, straw, and the boots of men tending animals affected with the disease also act as carriers of the contagion. The most infective and fatal medium is milk removed from infected cows and especially if given while warm. The period of incubation varies from two to five days, but we have known very young pigs to become stricken and die in eighteen hours, after receiving milk fresh from sick cows.

*Symptoms.*—Dulness, slight shivering, and a disposition to huddle together or burrow into the straw and manure are the first indications of ill-health. This is soon followed by lameness in two or all the limbs. If the feet be examined at the junction of skin and hoof it will be found red and inflamed, or small blisters will be seen which, sooner or later, break, and expose a red, raw-looking surface or superficial ulcer, which, by extending beneath the hoof, may cause it to slough. At this time vesicles usually also appear on the snout or lips or in the mouth, or they may be solely confined to the feet, or to one of the other parts named. Where the mouth is affected there is "slavering," and solid food is refused or taken sparingly. Fluids are eagerly drunk up as the result of the existing fever, which is denoted by a rise in the body temperature and other symptoms of general disturbance.

*Treatment.*—The patient is to be placed on clean dry litter and have the feet thoroughly cleansed. A saturated solution of alum with a little carbolic acid should then be applied to the ulcerated parts and repeated each morning and evening. This will have the effect of encouraging the wounds to heal and at the same time destroy the virus which escapes from them.

Applications to the feet may be made by means of a garden syringe or by driving the affected pigs through a shallow trough containing the solution. The affected animals should be kept in confinement and allowed to rest as much as possible. When at large the ulcerated feet become damaged and sloughing of the hoofs not infrequently ensues. Local applications to the eruption on the snout and other parts of the body may be made by the same means as that prescribed for the feet, care being taken not to allow more than a small quantity to enter the mouth.

A plentiful supply of good food in a soft condition is needed to keep up the animal's strength and prevent wasting. A little common salt and bicarbonate of potash may be mixed with the morning and evening meal. Where the bowels are constipated and the fever runs high, as may be known by the lowering ears, drooping tail, arched back, and unsteady movements, a small dose of sulphate of magnesia may be given either in the food or by draught, and repeated if necessary in twenty-four hours.

## ANTHRAX.

Anthrax is a specific contagious disease, to which all farm animals are more or less liable, and from which it may spread from one to the others either directly or indirectly. It is also communicable to man, rabbits, and various other creatures.

The virus, or contagion, consists of a minute rod-like organism termed *bacillus anthracis*, which on entering the blood undergoes rapid multiplication, and in addition to blocking up the minute vessels, gives out a poisonous principle destructive to life.

The reproductive power and virulence of this organism exceeds that of any other contagion affecting our farm animals. Once a few of these invisible particles gain access to the blood, countless thousands are quickly developed and a few hours is often sufficient to terminate a fatal sickness.

Their entrance into the body is no doubt mostly, if not always, effected through broken surfaces, the result of scratches and abrasions inflicted on the mouth during mastication, and other parts of the alimentary canal. They are not, as is sometimes supposed, given out in the breath, and like sheep pox and cattle plague, transmitted by atmospheric contamination.

In regard to pigs, the poison is mostly derived directly from the flesh and blood of animals having died of the disease, and the majority of outbreaks in hogs follow upon the death of cattle or horses whose parts or organs they have been allowed to eat in a raw condition.

Anthrax is very destructive, and seldom fails to kill the animal in whom the poison is allowed to enter.

*Symptoms.*—This disease invariably comes on suddenly, and runs a rapid course. Hence owners are frequently led to regard the outbreak as the result of malicious poisoning. Where the temperature of the body has been examined thirty-six to forty-eight hours after infection, a rise of from three to four degrees has usually been found; but at this time no other perceptible change may be noticed in the animal's condition. Very quickly, however, this is followed by sudden and extreme prostration : the ears, head and tail droop, and the animal wears a dull, dejected appearance. It is indisposed to move, and walks, when made to do so, with a rolling gait. Defective movement becomes most marked in the hind extremities, resulting in complete paralysis and inability to progress. The stricken beast may still raise himself on his fore limbs. Swelling of the throat and neck appears, and rapidly extends over the head and face, giving to the latter a bloated, hideous aspect. The breathing is now rendered difficult and the animal sits on his haunches with open mouth and protruded tongue, wheezing and gasping for breath, or lies prostrate and helpless.

Red congestive patches appear on the skin, and diarrhœa is not infrequently a prominent symptom. The duration of the disease varies from a few hours to two or three days.

*Treatment.*—In this, as in some other contagious affections, curative treatment offers no hope of success, and having regard to the fatality of the disease, and the great risk of multiplying and disseminating a contagion which may live for long periods in the soil and manure of our homesteads, it is to the interest of the stock owner to dispose of the stricken animal as speedily as possible. Measures of prevention against an extension of the disease and its re-appearance on some future day, are of the first importance.

In this connection the carcase should be well dressed with lime and immediately buried six feet deep in some place to which stock will not have access. In removing it to the place of burial care should be taken that no matter escape from the mouth and other outlets of the body. Without this precaution the poison or virus may be scattered along roads and pastures, and become a source of future trouble. On no account should the body be opened or blood caused to flow from it, unless this can be done in or against the grave. The reason will be obvious, when we repeat that every drop of blood swarms with the poison, and a minute fraction of that amount would be sufficient to destroy life. All the manure and litter contained in the sty should be burned, and the floors and walls washed with a strong solution of some disinfectant, twice at an interval of a week, and afterwards dressed with lime. Where affected pigs have had the run of open yards, all cattle should be removed from them at once and the manure taken out and heaped up in some ploughed field well mixed with lime and allowed to remain for three months before being used. Such manure should not be spread upon pasture land, nor should it be stored near to a watercourse.

## CONSTIPATION.

When the excrement is hard and wanting in moisture, it frequently happens that it is retained in the intestines for long periods and is with difficulty ejected from the bowels. This condition is what is known as constipation. The fault may arise out of some mechanical interference with the onward movement of the ingesta, as when a tumour presses upon the intestine, or by growing into it partially blocks the passage, or it may result from functional derangement of the bowel itself. In this latter connection the normal movement of the gut by which the food is propelled backward in the course of diges- tion may be weak and fail to effect its purpose, or the fluid secretions which ordinarily keep the intestinal contents in a soft condition may be insufficient in quantity or altered in quality, and in consequence of one or more of these causes the excrement is retained and allowed to accumulate until the bowels become unduly distended, and their contents with difficulty discharged.

*Causes.*—The causes of constipation are many and various. The chief of them are overfeeding, insufficient supply of water, the long-continued use of astringent articles of diet, such as ripe acorns, beech mast, and chestnuts. General debility and insufficient formation of bile are now and again determining causes. Ravenous consumption of dry indigestible food after a long fast will sometimes induce it. Constipation attends the early stages of swine plague and invariably results from poisoning by lead. Large heavy pigs and sows advanced in pregnancy, who take little or no exercise while receiving an unlimited amount of concentrated food, are specially liable to this disorder.

*Treatment.*—It is important to remember that in this disease the alimentary canal is overburdened with matter, and for this reason the supply of food should, for a time, be much restricted or altogether withheld.

The withdrawal of solid aliment for twenty-four hours and the substitution of simple wash, with a small allowance of roots, will lighten the burden of the bowels and assist in restoring their normal activity. A bold dose of sulphur and sulphate of magnesia may then be administered in a little tempting food, or should this be refused castor oil may be offered in the same way; where both are declined one of them must be forcibly but carefully administered, and again repeated if the bowels do not respond in twenty-four hours.

Injections of warm soap and water will prove serviceable in relieving the posterior bowel, and should be administered once or twice a day until a free action is induced.

If there be a tendency to return of the constipation after once the bowels have been relieved, a little common salt and sulphur with powdered nux vomica should be given in the food for three or four days. Where habitual constipation exists the most reliable corrective will be found in a liberal daily ration of roots or a little linseed oil mixed with the ordinary food morning and evening. In heavy breeding sows with costive habit one or the other should always be provided where a run of grass is not available.

## DIARRHŒA (SCOUR).

The term diarrhœa is used to express an irritable state of the bowels in which the dung or fæcal evacuations are expelled in a fluid or semi-fluid state, without exhibiting the severe griping and straining that attends certain forms of inflammation.

*Causes.*—Pigs endure some conditions which readily provoke diarrhœa in other animals. Of these we may mention foul water and decaying animal and vegetable matter, which, to a certain extent, they take with impunity. Diarrhœa usually results from the direct action of some irritant contained in the food or water, or it may arise out of a vitiated state of the blood.

The liability to diarrhœa is much greater in young pigs than in older ones, and especially during the sucking period. The particular causes of the disease are sudden change of food, as when rank rapidly-grown grass or certain acrid plants are taken after the continued use of more substantial fare. Intestinal worms and the excessive consumption of putrid animal and vegetable matter, the indiscriminate administration of salt, either as brine or otherwise, and functional and structural diseases of the liver, are the more frequent causes of the disease. It also occurs in pulmonary tuberculosis, rickets, and some other constitutional affections. In the first-named disease it invariably assumes a chronic form and continues throughout life. In sucking pigs it must be sought for in some vitiated state of the milk of the dam, or insanitary state of the stye, and not unfrequently also in decomposing matter fouling the teats of the sow.

*Symptoms.*—These vary somewhat considerably in different cases. Sometimes they only amount to a transitory looseness which quickly passes away without giving rise to pain or other signs of illness. In others the discharges continue to be thrown out for days; there is then more or less wasting and loss of appetite, with coldness of the skin and extremities and general dulness and depression. Where diarrhœa is accompanied with pain the back becomes arched, the belly tucked up, and the pig is restless. The excrement is offensive in odour, and sometimes contains food in a crude or half-digested state.

In young suckers the discharges are sour-smelling, mixed with mucus, and contain solid masses of curd and may be streaks of blood.

*Treatment.*—No hard and fast rule can be laid down for the treatment of diarrhœa. The measures to be adopted will depend upon the cause giving rise to it, and the stage and indications of the attack. The first step to be taken is to remove the cause, if that be known, and generally it is good practice to cleanse the bowels of any offensive matter they may contain by a dose of castor oil, given in a little warm milk, or if refused in this form it should be administered as a drench. Where indications of pain are presented a little powdered opium with catechu, bicarbonate of potash and precipitated chalk may be given in the food two or three times a day. The diet should be very simple, and may consist of well boiled cake gruel, linseed tea, or wheat flour boiled with a little carbonate of soda.

In the case of diarrhœa affecting suckers the sty should be kept in a thoroughly clean condition, and supplied with a liberal amount of clean straw. They should, moreover, be protected from cold and wet. The food of the dam should be sound and wholesome. Some young sensitive sows are so irritated by the sharp teeth of their offspring that their milk undergoes change and acts prejudically on the young. Where this is so the little tusks must be nipped off.

## GASTRITIS.

Inflammation of the stomach, or, as it is technically termed, gastritis, is of frequent occurrence in the pig, although during life it is seldom recognised as such. The liability to this disease is no doubt connected with his peculiar scavenging habits, which, under circumstances of hunger, lead to the ingestion of various irritating, and even poisonous substances. It is further increased by the exceptionally insanitary conditions under which the food of this animal is provided, and the many sources of contamination to which it stands exposed.

*Causes.*—Among the causes of acute gastritis the most common are salt and saltpetre. In the form of brine these two substances frequently find their way into the "pig tub" by being thoughtlessly mixed with the "wash." So long as the latter is plentiful and ample dilution takes place the former may prove harmless, but where wash is present only in small amount, and brine is given in a concentrated form, gastritis frequently arises in consequence. Arsenic and some other mineral poisons existing in "sheep dip" and wheat dressing have on numerous occasions been known to induce it. Common washing soda, after being used for domestic purposes, sometimes gains access to the " wash " in dangerous amount.

Over-feeding after a long fast, and especially if the food be of an indigestible nature, may provoke an attack, as may also the consumption of large quantities of raw putrid flesh and other garbage.

It sometimes occurs that where pigs are fed on oats, the small hairs on the grain will aggregate together and form large oat-hair concretions. These may reach the dimensions of a cocoa-nut or even larger, and by their presence mechanically irritate and inflame the stomach. Several cases have been brought to our notice in which *post-mortem* examination has shown death to have resulted from this cause.

Inflammation of the stomach is frequently observed in swine plague, and in chronic heart disease of an obstructive nature it commonly appears in a sub-acute form.

*Symptoms.*—A disinclination to feed and a pinched and anxious expression of the face are the first signs of suffering. The patient isolates itself from the rest of the herd and refuses to come up to the troughs at feeding time. The body is tucked up and the back arched. There is considerable restlessness and a shifting from place to place, with a subdued grunt or squeal. As the disease advances the skin becomes hot and red, and there is great thirst. If solid food be taken the symptoms are aggravated, and colicky pains, with vomiting and retching, are excited. After the stomach has been emptied in this way the animal appears relieved, but has at the same time a great desire for water.

The action of the bowels varies in different cases; sometimes they are constipated, the fæces being hard and pellety, at others, where the inflammation has extended to the bowels, there is violent diarrhœa and straining. Where vomition does not occur the belly becomes distended with gas, and pain is excited by deep pressure over the region of the stomach.

*Treatment.*—It is of the first consideration in dealing with gastritis that the stomach be put to rest. No solid food should therefore be offered, but small quantities of bland nourishing fluids may be given from time to time. Of these, linseed tea or very thin well-boiled oatmeal or flour gruel, given cold, are the most suitable. If there be reason to suspect the existence of constipation, a purgative, in the form of a little castor oil, will be required to empty the bowels, and should the pain be severe a small dose of opium may be administered at intervals of two hours. Distension of the belly with gas —the result of fermentation—must be met by a few drops of carbolic, mixed with a little glycerine, given in linseed tea.

When the acute symptoms subside there should be no hurry to return to the use of solid food. A little milk and, if

10

it can be obtained, fresh bullock's blood well mixed together, should be given for the first day or two, to be followed by repeated small rations of well boiled oatmeal and sharps, and so soon as convalescence is re-established the ordinary diet may be resumed, but it must be given sparingly with a liberal supply of water or milk.

### INTESTINAL WORMS.

Having regard to his scavenging habits and the insanitary conditions of his environment, the pig may be said to suffer comparatively little from intestinal worms.

In this country the parasites which take up their abode in the bowels of this animal are probably few in number, and of these we know of only one that can be regarded as of common occurrence. This, the *Ascaris suilla*, is a nematode or round worm of a milky-white colour, thick in the middle and taper-ing towards each extremity. It is several inches in length, and has a firm, elastic body, and small head with three denticulated lips.

This worm infests the small intestine, and we have seen it to exist in such numbers as to completely fill the gut for a considerable portion of its length.

*Symptoms.*—The *Ascaris suilla* is a chyle-feeding parasite, *i.e.*, it lives on the alimentary principles found in the bowels of its host, and when present in large numbers starves the latter by reducing its supply of nourishment. To compensate for this double demand a much larger quantity of food is con-sumed. In spite, however, of an increased supply the animal loses flesh and becomes weak and impoverished. The skin is dry, scaly, and "hide bound." The back is arched upward, and the belly is either closely tucked up in the flank, or it is unduly large and let down. The latter condition usually prevails when the appetite is ravenous, and is commonly spoken of as "podge-bellied." Pigs largely infested are fretful and unsettled; they wander about grunting and squealing, and seldom rest as do healthy stock. Sickness, during which one or more worms are ejected from the stomach, is sometimes present, and fits of a convulsive or epileptic nature are, in some cases, frequently seen. Irregularity of the bowels, with occasional diarrhœa, appear in the later stages of the disorder, and the appetite falls away, thus adding to the emaciation and weakness.

*Treatment.*—Newman recommends about 2 drachms of decorticated castor oil seed mixed with the food. Should this fail, sanatonine or male fern may be tried. Whichever remedy is selected it should be given after fasting and followed by an aperient dose of Epsom salts. The strength in these cases should be well supported by a liberal allowance of good food, to which a little salt may be added at each meal.

## MEASLES.

This is a parasitic disease, due to the presence of small " bladder worms " or " measles " in the substance of the muscles. They are termed bladder worms on account of the resemblance they bear to little sacs or bladders. Technically, they are known as *Cysticercus cellulosæ*. These organisms, when consumed by man, as they sometimes are with pork, develop into tape-worms, and should the eggs of the tape-worm find their way back into the pig, they in turn will give out bladder-worms. The young bladder-worms, when hatched, bore their way through the bowels of their host, and ultimately reach the muscles, where they attain their full development.

The measle, or bladder-worm, therefore, is to be regarded as a young tape-worm, requiring to pass from the pig into the bowels of the human subject in order to complete its development. Measly pork, it will be seen, is one means of conveying tape-worms to man, and should not therefore be used as human food unless thoroughly cooked.

It is very difficult, and as a rule impossible, to recognise the presence of measles during life. Pigs infested by them suffer in various ways, depending upon the organ in which the parasite is located.  In many cases they fatten, notwithstanding the presence of large numbers, which are only recognised when the carcase is cut up.  We have seen the liver studded with them throughout, and the flesh to contain a considerable number without ill-health being induced, and we have known half a dozen, when located in the brain, to occasion violent fits and death.  When large numbers infest the muscles, the animal moves stiffly and shows signs of soreness.  The body wastes and the appetite falls off.  There is general dulness and indisposition to move.

Should the liver be extensively affected at the same time, these symptoms are aggravated and the bowels become irregular, being at one time constipated and at another loose.  The skin may also become yellow, owing to functional disturbance of the liver.

Nothing can be done in this affection in the way of cure.  The parasites are embedded in the substance of the flesh, and cannot be destroyed by any known medicine.  Where they do not kill the patient, they will in time die, and later on become changed into little white gritty particles.  To prevent it pigs must not have access to the excrement of man in which tapeworm exists.

Fortunately the disease is not often seen in this country, but in Germany and some other parts of Western Europe, where pork is consumed in a partially raw state, tape-worm in man and measles in the pig are common ailments.

### PROTRUSION OF THE RECTUM.

That part of the bowel which ends with the anus is known as the rectum, and it sometimes happens that a portion of it is forced out and projects beneath the tail in the form of a red fleshy-looking " tumour."  This condition constitutes prolapse or protrusion of the rectum.

Animals who are debilitated, or extremely fat, or pregnant,

are most liable so to suffer. The exciting causes are violent straining, such as takes place in constipation of the bowels, difficult farrowing and obstruction to the escape of urine, by "gravel," or other forms of interference.

Protrusion of the rectum sometimes results from diarrhœa, and is also provoked by the presence of intestinal parasites.

The protruded gut may not be more than half an inch in length or it may be several inches. At first it is of a bright red colour and covered with a sticky mucus. After exposure for some time it darkens and assumes a livid or bluish red appearance. If much protruded it becomes swollen, tense, and may crack and bleed.

*Treatment.*—Protrusion of the rectum is sometimes difficult to deal with in pigs owing to their refractory nature, and when the patient is fairly fat and fit for the butcher it should be destroyed. In the case of animals required for exhibition this may not be desirable, and therefore treatment must be attempted. In this connection there are two points to be dealt with, first, to return the gut, and secondly, to keep it in its place.

Before attempting replacement, the protruding part should be thoroughly cleansed, and if, as sometimes occurs, it is black and cold and swollen, fomentation with warm water slightly carbolised is desirable as a preliminary step to its return. If the animal strains, a little opium should be administered, or a dose of morphia may be mixed with a little treacle and put in the mouth. The protruded gut, well oiled, is then to be seized with the left hand and while being thus steadied it is pushed gently back with the oiled finger of the right. The animal is now to be kept without solid food for from twenty-four to forty-eight hours. Should it be forced out again the exposed gut must be again dressed with carbolised oil and returned as before. Where constipation is the inducing cause a dose of castor oil must be given to freely open the bowels, and any straining that may exist is to be subdued by a dose of opium or morphia given in the wash.

Syringing the anus with cold water will sometimes prevent a second protrusion of the part.

## HERNIA (RUPTURE, BURST).

By hernia is understood the escape of some organ or part of an organ from its proper cavity, as when a portion of intestine passes through a rent in the muscles of the belly and is contained in a pouch beneath the skin. This displacement of parts may take place either through an opening accidentally made, such as that already referred to, or through a natural orifice.

The forms of rupture most frequently seen in pigs are those known as scrotal hernia, and umbilical hernia—the former consisting of the passage of some part of the abdominal contents into the scrotum or purse with the testicle, the latter being an escape of the same through the imperfectly closed navel opening. Both these varieties of rupture take place in early life, and are, therefore, generally seen in young pigs.

*Causes.*—The predisposition to hernia is hereditary. Rupture may occur in any part of the wall of the belly as the result of a blow; it may also arise from straining, as during parturition and in the forcible expulsion of urine and fæces where obstruction to their passage exists. Slipping, jumping, and violent exertion of any kind may give rise to it. Scrotal hernia is mostly due to congenital defect; the opening through which the testicle passes from the belly into the purse being abnormally large, thus allowing the gut to escape.

*Symptoms.*—In the examples referred to, rupture is recognised as a soft, more or less fluctuating enlargement, standing out from the wall of the abdomen. In some instances it appears quite suddenly, in others it is slowly developed. One marked feature of these swellings is the variation in size which they undergo from time to time. After feeding there is a notable increase and a corresponding diminution after fasting. When pressed the swelling may give out a gurgling sound.

This arises out of the displacement of air in the escaped intestine.   By a little careful handling the contents of the hernia can generally be forced back into the belly, and the opening by which they escaped may then be felt.

*Treatment.*—Ruptured pigs should be fatted and disposed of as speedily as possible, as the conditions out of which the disease arises are usually hereditary.   Sows and boars throwing offspring of this kind should not be used for breeding purposes, nor should any part of a litter where two or more are affected be bred from.

### APOPLEXY.

Apoplexy is a disease of the brain in which one or all of three conditions may exist.   (1) The vessels may be unduly charged with blood (congested) ; (2) some of them may have broken, and allowed their contents to escape (hæmorrhage); (3) congestion, with escape of the watery parts of the blood (serous effusion).

Pigs are specially liable to apoplexy, and there are few farmers who have not experienced loss from it at one time or another.   Owing to the suddenness of its onset and the rapid and fatal course which it runs, it is often mistaken for poisoning, and the contents of the stomach are submitted to chemical analysis, when the mischief should be sought for in the contents of the head.

*Causes.*—Apoplexy is mostly seen in young pigs in close confinement, and after being suddenly transferred from moderate to rich and abundant fare.   Forcing with animal food, such as the blood and refuse of slaughter-houses, after a scant farinaceous or vegetable diet, is perhaps the most common inducing cause.   It also results from sudden exertion on the part of very fat animals.   Heart disease and certain forms of kidney disease will induce it, and it sometimes arises from over-distension of the stomach with solid food.   Violent straining during parturition, especially in old sows, is also an occasional, and perhaps the most uniformly fatal cause.

*Symptoms.*—The most striking feature of apoplexy is the suddenness of its onset and the rapidity with which it destroys life. In the pig there are seldom any warning symptoms of its on-coming. Animals seemingly in robust health are seized with a fit and fall to the ground as if "shot down," or this may follow upon a brief convulsion, in which the body shakes, and staggers helplessly over. There is no effort to rise, and all the limbs are limp, and helplessly paralysed. When raised and again liberated they fall "like a stone." At the moment of the attack consciousness disappears. The vessels of the head are engorged with blood, and the lips and snout and tongue become dark red in colour, and later blue and livid. The eyes are widely opened, and may be touched without exciting movement of the eyelids. The breathing is slow and deep, and accompanied by a fluttering or snoring noise. Urine passes from the bladder involuntarily, and excrement from the bowels. No movement is excited by pricking the skin with a pin, all sensibility having been destroyed by suspension of brain function.

The duration of the disease varies from a few minutes to a few hours, and recovery is seldom effected.

*Treatment.*—But little can be expected from treatment in a disease so sudden and severe, and considerations of humanity no less than economy, dictate immediate slaughter. The flesh of animals destroyed while suffering from apoplexy is perfectly wholesome and may safely be used for human consumption, after prompt bleeding, provided no other disease inimical to public health exists.


### Epilepsy—Falling Sickness.

Epilepsy, like apoplexy, is a disease of the nervous system. They still further resemble each other in so far that they both occur suddenly in what are known as fits. Their points of difference, however, are very striking, for while, as we have seen, the latter is almost always fatal, the former is seldom so. Epileptic fits may occur again and again at varying intervals for weeks, months and years without seriously damaging the general health of the patient. Young animals are most frequently its victims, for reasons which will presently appear, but older ones are by no means exempt from it.

*Causes.*—Epilepsy is the result of a morbid irritation affecting some part of the nervous system. Blows on the head, tumours in and upon the brain, fright and excitement occasionally induce it; engorgement of the stomach, particularly after a long fast; but worms in the intestine and the irritation resulting from teething are its most frequent causes. We have also seen it to result from the presence of gravel in the bladder, oat-hair concretions in the stomach, and calculi in the kidney. Excitable young boars and yelts about the time of puberty sometimes suffer from it, as do also sows when impoverished and weakened by excessive suckling.

*Symptoms.*—As a rule, to which there are few exceptions, epilepsy is sudden in attack and comes on without warning symptoms. As in apoplexy, the animal falls to the ground in an unconscious state, but while in the last-named disease the body and limbs are limp and motionless, in the former the entire frame is more or less violently convulsed. After falling to the ground the legs are extended and the muscles contracted so that the body becomes stiff, rigid and tremulous. The jaws are closely clenched, froth issues from the mouth, and the eyes roll in a wild stare. For a few seconds or minutes the legs are moved involuntarily forward and backward in a fighting manner. Then the jaws relax and a few champing movements are made. The convulsions then cease, the muscles of the body relax, and consciousness returns. The animal now lifts its head with a vacant look and soon gets up. For a few paces its movements are unsteady and rolling, as the result of exhaustion, but this soon passes away, and in a short time the beast appears as if nothing had happened.

In some instances the return to consciousness is followed by symptoms of excitement or even frenzy, in which case the animal squeals, jumps about, runs with its head against the wall, and behaves in a frantic manner. Such attacks are followed by extreme prostration which may continue for several days.

The duration of the attack varies from time to time. It may not be more than two or three minutes, or it may extend over ten or twenty, or even half an hour. Sometimes one attack is quickly followed by another and another, and this may go on for a succession of attacks extending over many hours. Now and again these cases end in fatal exhaustion.

*Treatment.*—At the outset of the fit the pig should be placed on its side. A piece of stick thrust across the mouth will enable those present to pull the tongue forward. This should be performed quickly, or suffocation may result by its falling backward over the air passage. A little cold water may be dashed over the face now and again, but beyond this there is little else to be done while the fit is on. Much more importance is to be attached to measures of prevention. In this connection we have to consider the causes by which the disease is induced. Of these, parasites we have said are one of the most fruitful, and especially the large round worm (*Ascaris suilla*) which infests the small intestine. When pigs are known to harbour these creatures, a dose of santonine, or the extract of the root of the male shield-fern should be given in milk after 12 hours' fasting, to be followed by a dose of castor oil 4 hours later.

When the disease appears in sucking pigs, either from derangement of the mother's milk or from teething, the bowels of the dam should be freely opened by epsom salts and sulphur, and it is also important that the food should be carefully examined, and anything calculated to render the milk deleterious must be discontinued. Its appearance in the sow must be met by more and better food, and the removal of the young. In all instances an entire change of food may be made with advantage.

## Tetanus.

Recent investigations into the nature and origin of this disease have clearly shown it to be a contagious affection. The organism to which it is due has been found to exist in the earth of our gardens and fields, and we have ample evidence in our own experience of its power to spread from one animal to another where wounds exist and afford an opportunity for the entrance of the virus into the system. The microbe or organism of tetanus does not, like that of anthrax, infest the blood, or enter the flesh of remote organs as in tuberculosis, but remains near the wound by which it enters, and there gives out a poison whose action on the nerve centres results in the peculiar cramp or tetanic state of the muscles. The disease is most common after castration and during the changing of the teeth, but the poison when present may enter the body through any surface wound.

*Symptoms.*—The incubation period of tetanus is about two to four days, after which the pig is noticed to move stiffly. The head is protruded, the ears stand up in a stiff manner, the eyes stare, the face wears a pinched and anxious expression, the breathing is short and panting, and the belly is tucked up. When made to turn the body is found to be stiff and the movements unsteady and staggering. The back may be arched upward or downward. Ultimately the patient becomes quite rigid and falls exhausted and helpless and lies with the legs stiffly extended.

Pigs affected with this malady rarely recover, and should therefore be destroyed as soon as its existence is recognised. The flesh around the wound should be removed and burned, and the rest of the carcase may be used for food without scruple.

## Rickets.

Rickets, although specially manifested as a disease of the bones, must be regarded as a general or constitutional affection in which the nutrition of the body as a whole is impaired and its growth and development retarded.

It is a disease of early life, occurring during that period
when growth is usually most active, viz., from a few weeks to
a few months old.  Rickets is undoubtedly hereditary, and
has been known by us on several occasions, both in pigs and
other farm animals, to re-appear in the offspring of affected
parents.

It is most frequently seen in the progeny of old, poor and
ill-conditioned sows, which have been too freely used for
breeding purposes, and we have known it to occur in the young
of sows fatted for show purposes.  Confinement in small, wet,
stuffy sties, with bad living and exposure, is believed to act as
an exciting cause, but we must confess to have seen it in model
sties and under ideal management.  Where predisposition is
strongly inherited the exciting causes often escape notice
altogether on account of their simplicity.

*Symptoms.*—The early symptoms of this affection are mostly
overlooked, and it is not until deformities of the limbs appear
that the real nature of the disease is recognised.  Slow growth
and general unthriftiness of the affected pigs are the first signs
of ill-health, and careful observation may detect an occasional looseness of the bowels with sometimes diarrhœa, in
which the fæces are pale and sour-smelling.  The appetite is
uncertain and never so vigorous as the healthy portion of the
herd.  Coupled with a general appearance of unthriftiness
there is a great desire for warmth and an indisposition to leave
the bed when the other animals are out and about.

Soon there appears some stiffness in movement, which is
usually attributed to cramp or rheumatism, and actual
lameness in one or more limbs may appear later on.  At this
time the joints are observed to enlarge and to show tenderness
when pressed.  The legs appear short, and become more or
less bowed either inward or outward.  The body now becomes
nearer the ground, and the animal has a diminutive dwarfed
appearance.  When the disease involves the spine various
forms and degrees of curvature are observed.  Sometimes the
back sinks and presents a hollow appearance, at others it is

arched upward or "roached." Sometimes the deformity is confined to the limbs, at others to the backbone, while in a third both may be more or less affected. The bending and other alterations in the shape of the bones are due to a soft and yielding state, partly arising out of a deficiency of earthy matter, and partly also from excessive and too rapid growth of the animal constituent. In some exceptional cases the changes in the bones are not preceded by any premonitory signs of illness, but gradually develop while the animal still thrives and even becomes fat. These cases are invariably slow in their progress and much less severe than those previously noticed.

*Treatment.*—Ricketty pigs do not generally prove remunerative to keep, although in the majority of cases the time comes round when the general health becomes restored, the bones acquire their natural hardness, and growth is re-established ; but they never reach the size of their fellows. The first thing to do is to uphold as far as possible the general health.

The sty should be dry and spacious, fairly warm without being close, well ventilated, and sheltered from the east and north. A liberal supply of clean litter should be given, but no opportunity should be afforded them to bury themselves in manure. The food should be carefully selected and prepared for a time until the bone softening ceases. Cooked food, such as boiled potatoes, bran and barley meal, mixed with a fair proportion of ground oats and well-boiled pea-flour, is the most suitable aliment. A small quantity of cod-liver oil may be added in the morning and a little sulphate of iron in the evening. Cinder ashes and wood charcoal should be allowed for the animals to pick over, and some benefit may be obtained by adding a little limewater or calcined bone to the food or putting a few bits of chalk in the pen.

Pigs having suffered from rickets should not be kept for breeding purposes.

## RHEUMATISM.

Rheumatism, frequently described as " cramp," is a constitutional disorder sometimes assuming the form of a fever. It is characterised by stiffness, or more commonly lameness, which may or may not be attended by inflammation and swelling of the affected parts. It varies considerably in severity and duration in different cases and in the same animal at different times. It may be acute and quickly fatal, but it mostly assumes a chronic character and continues for long periods, or disappears and returns again at longer or shorter intervals. Besides the limbs and trunk, it sometimes attacks the heart and proves fatal by damaging and disarranging its valves.

*Causes.*—The pre-disposition to rheumatism is hereditary ; that is to say, animals born of parents having suffered from the disease inherit a liability to suffer from it themselves when exposed to exciting causes. Old animals and young animals appear to be more liable to it than others in middle life. Certain localities, as low, cold, undrained flats predispose to the disease, and cold, damp sties. Spring and autumn are the periods of the year when it mostly prevails.

The exciting causes are chiefly chills from over-heating, long exposure to cold and wet after fatigue and fasting. Sows when suckling often contract it if cramped by confinement in small, damp, stuffy sties, and there is some reason to believe that disordered digestion is in some way conducive to it. Sprains and injuries to the legs and back will sometimes provoke an attack of rheumatism in susceptible subjects.

*Symptoms.*—Manifestation of stiffness, pain and lameness are among the more prominent symptoms of the disease, but it is the changeable character of these that mainly distinguishes it from ailments the result of accidental causes.

Rheumatism usually comes on suddenly, at first presenting itself as a mere stiffness in one or all the limbs, or along the back, or across the loins—the latter being known as lumbago or lumbar rheumatism.

In other cases it commences as severe lameness, or excru-
ciating pain attended with more or less swelling of the joints.
From one limb it suddenly passes to another, and another,
until all may become affected, or it will suddenly leave one leg
and attack its fellow on the opposite side. The same variation is noticed from day to day in the severity of the disease.
One day the patient is much better and worse on the next.
It is this shifting and variable character of the malady that
distinguishes rheumatism from other disorders. The swollen
joints and other structures implicated are hot and painful,
with a tendency to hardness. They frequently remain permanent after continuing for a certain period.

Animals affected with rheumatism lose flesh, the skin
becomes rough, scaly, and unthrifty, the appetite fails, and
they lie about with no disposition to move. If the disease is
acute, the temperature of the body is raised, and the heart's
action is quick and jerky, and may be intermittent. The
symptoms are aggravated by cold easterly and north-easterly
winds and a damp atmosphere, and become less severe under
dryness and warmth.

*Treatment.*—In commencing to combat this troublesome
affection it is of the first importance that the patient be
placed in a fairly warm and dry apartment, free from draught,
with an abundant supply of clean bedding. The bowels are
then to be freely opened by a full dose of Epsom salts,
and this must be repeated in four or five days' time, and again
after a similar period if no abatement in the symptoms is
brought about. A dose of iodide of potassium in the morning
and of bicarbonate of potash in the evening may then be given
from day to day until some perceptible improvement is
observed.

When the pain is severe relief will be obtained by giving
the animal a small dose of opium every eight or twelve hours,
according to the necessities of the case. Where local treatment is resorted to, the application of a liniment composed
of belladonna extract and glycerine, and warm dry bandages
to the affected part is all that is required.

The diet should be strictly regulated both as to quantity and to kind. Barley meal, peas, and maize should be avoided. Oatmeal, crushed oats, pollard, with ground linseed cake and a few roots form the best and most suitable diet. Preventive measures require that animals once affected with rheumatism shall be cast, and no longer used for breeding purposes. The sties should be well drained, sheltered from the north and east, efficiently ventilated, and kept dry and clean on the ground surface.

### GRAVEL.

The urine of all animals contains a certain varying amount of salts in solution. These, for the most part, are derived from the tissues and represent the waste products of physiological work. The chemical composition of urinary salts varies in different animals, and is largely influenced by the nature of the food on which they live. In the pig these salts consist chiefly of the phosphates of ammonia and magnesia, with oxalate of lime. Under ordinary circumstances they are discharged with the urine, but it sometimes happens that they are deposited in the bladder instead, and accumulate there in large amount. We have on several occasions seen this organ filled with them. When present they have the appearance of a white powder, not unlike finely powdered sugar.

The cause of this accumulation is in great part due to the long retention of urine in the bladder. Very fat pigs often lie for long periods without urinating, thus allowing the salts to crystallise out and deposit ; or the bladder may be paralysed and unable to expel its contents, or some obstruction to their discharge may exist in the neck of the bladder or in the course of the penis. Any of these faults—causing retention of urine —favours the deposition of " gravel." Male animals are more liable to gravel than females.

*Symptoms.*—When any substantial amount is present, the patient stales frequently, and in small quantities, or the urine may be discharged involuntarily, or be altogether suppressed ;

or the stream is suddenly stopped once or oftener while the urine is being expelled. The colour of the fluid may be milky, owing to the presence of salts. In severe cases the end of the penis may swell, and the organ protrude, but the most certain indication of the presence of gravel is obtained by feeling the bladder with the finger passed up the rectum.

*Treatment.*—Animals suffering from this disease should be prepared for the butcher and slaughtered.

As a preventive fat pigs should be allowed a little exercise daily, or be driven round the sty once or twice at each time of feeding, so that they may be induced to empty the bladder.

## URTICARIA (NETTLE RASH).

In this disease the skin becomes congested in streaks or patches and the watery constituents of the blood ooze from the vessels and give rise to appearances similar to those produced on the human hand by the poison of the common stinging nettle.

*Causes.*—Some animals are specially disposed to urticaria on account of a morbid sensibility of the skin; for the most part, however, it is the result of derangement of the stomach and bowels, or a congested condition of the liver brought on by errors of diet. The excessive use of peas, and barley meal or new beans, and acorns have been known to occasion it, and we have seen it occur from the use of decayed roots and potatoes. It may also be induced by lice and by violent scratching or rubbing an over-sensitive skin. Young sows during the first pregnancy sometimes suffer severely from it.

*Symptoms.*—The eruption usually comes on suddenly after a little constipation, or diarrhœa and dulness. It consists of white raised patches surrounded by a red congested band of skin. In some cases it is attended with a good deal of itching, while in others this is only slight or altogether absent. The patches may be few in number and not larger than a sixpence, or they may be scattered over the body and reach the size

11

of a crown piece.  One of the peculiar features of the disease
is the suddenness with which it appears and disappears without leaving any trace behind.  It is seldom attended with
any marked indications of illness.

*Treatment.*—In this affection the stomach and bowels need
rest, and the over-loaded vessels of the liver require to be
relieved.  This may be effected by an active purge and the
withdrawal of all food for twelve hours.  A dose of castor oil
in milk may be given and repeated in a day or two should the
eruption continue or return.  This may be followed by a dose
of bicarbonate of potash and common salt night and morning
for a week, during which time the diet should be carefully
regulated, both as to quantity and quality.  A little exercise
and a bite of grass will prove beneficial where it can be
obtained.

### ECZEMA.

Eczema is an eruptive disease of the skin occurring in
scattered patches of redness in which small groups of vesicles
or blisters appear as the result of a circumscribed inflammation of the superficial parts of the skin.

*Causes.*—Young animals and old ones are most liable to
suffer from this disease.  The exciting causes are mainly
concerned with food.  Sudden changes from a poor, indifferent diet to luxurious living, the too free use of animal
offal—as horseflesh, blood, and slaughter-house refuse—are
common causes of the disorder, and especially when given
in a state of decomposition; also sudden changes from one
description of food to another, as when animals pass from a
vegetable or farinaceous diet to one largely made up of animal
matter.  New beans gathered from stubbles after harvest
sometimes excite it in young pigs, especially if the weather
is hot and the animals are exposed to a burning sun.  It is
often a result of intestinal parasites, and in old animals chronic
diseases of the kidneys and liver are common inducing causes.
Poverty and high living may likewise occasion it.

*Symptoms.*—Eczema generally comes on suddenly. The skin presents a number of rounded patches, somewhat raised and red in colour, from which a watery fluid oozes. The parts exhibit an inflamed and blistered appearance at first, after which the moisture dries up and leaves the skin scaly or even scabby. A little later the scabs fall away and a slightly reddened surface remains. Where it is attended by much irritation the scabs are soon rubbed off and bleeding sores appear scattered over the skin. The eruption may cease altogether in a few days, without medical treatment, or it may continue to be thrown out on different parts of the body from time to time for several weeks, and in old animals for months.

*Treatment.*—The bowels should be freely opened at once. Castor oil, or a combination of sulphate of magnesia and sulphur will be found to answer the purpose.

A change of food is most desirable in these cases, and where the animals are closely confined they should be allowed to run at large. If possible a turn-out in an orchard or grass field for an hour or two in the course of the day should be allowed. After the aperient has ceased to operate, a little gentian powder and carbonate of soda may be given night and morning until the eruption ceases.

When it occurs in sucking pigs, as is sometimes the case, the food of the sow should be changed and the prescribed treatment carried out on her, so that the faulty quality of her milk may be corrected. Where the body of the affected animal is dirty it should be carefully washed in tepid water containing a small quantity of sulphur, and the affected parts of the skin may be dressed with zinc ointment. A dry, clean sty, with a plentiful supply of short litter, is much to be desired.

## SCABIES (ITCH).

Although pigs frequently show signs of irritation of the skin by scratching and rubbing, true itch, or, as it is technically termed, *scabies*, is very rare in the pig herds of this country.

It is not unfrequently seen in Holland, and in the centre and west of France.* Scabies of the pig, like scab in sheep, is a parasitic affection due to the presence of a minute parasite or acarus, termed *sarcophes scabiei*, and it has been said to be contagious for man; but however that may be, it is pretty certain that the pig sarkopt will not live long on the human body.

*Symptoms.*—As observed by Neuman, scabies begins with a violent itching, and appears to be at first confined to the head —chiefly on the ears and around the eyes, then it attacks the withers and croup, and inner surface of the thighs; later, it invades the entire surface of the body. Small red points, like flea-bites, are the first noticeable changes on the skin, then little pimples rise from the inflamed surface, and out of them a watery fluid exudes, which gradually becomes thicker, and partakes of the character of pus (matter). This sooner or later dries into a scab and either falls off or is dislodged. The constant rubbing to which the skin is subjected adds considerably to the mischief, by converting the papular eruption into open sores and superficial abrasions. The bristles fall out of their follicles, and the skin thickens and cracks at and about the parts infested with the parasite. In some situations wart-like excrescences spring up beneath the scabs, and when rubbed bleed freely. If not checked the disease continues to spread from day to day, until the entire body becomes covered over with a scabby eruption, with here and there cracks and sores. The incessant irritation materially affects the general health of pigs suffering from this disease and they consequently fall away in flesh, and, if neglected, die from exhaustion.

*Treatment.*—One of the many dips used for sheep scab may also be tried here, or a decoction made by boiling six ounces of tobacco in as many quarts of water. The old remedy, consisting of a mixture of linseed oil and sulphur with a little oil of tar, will also be found an efficient dressing.

---

* Neuman, by Fleming.

Whatever is used it should be repeated every twelve days for three times. The sties should, at the time, be thoroughly cleansed, and the manure and litter removed from them burnt.

The walls and fittings will require to be dressed with a strong anti-parasitic solution, such as arsenic, carbolic acid, or zinc chloride.

## Sore Teats.

Sows with delicate skin, and especially yelts, frequently suffer from sore teats. This may be due to an eruption (eczema) or to chapping, or it may arise out of the irritation induced by the sharp tusks of the suckers. Where it results from the last-named cause, the offending teeth should be nipped off with a small pair of sharp pincers. Mr. Sanders Spencer, in the management of his herd, frequently adopts this course, to the mutual advantage of dam and offspring. The former rests and gives up her milk freely to the young, and the latter thrive and grow much better than when their little needle-like teeth are retained intact. Where soreness is due to eczema, or chapping, a small dose of castor oil may be given to the sow, and the teats are to be sponged two or three times a day with a solution of borax and glycerine.

# CHAPTER XIV.

## BACON AND HAM CURING.

THE modern process of bacon curing differs essentially from what was in use up to a few years since. The demand of the public is almost entirely for lean meats of mild flavour, and, as a natural consequence, bacon curers have been compelled to alter the old methods. Scarcely a factory exists at the present day which has not a complete complement of machinery and mechanical appliances adapted to one or other of the many operations necessary in the expeditious output of cured bacon. It does not follow, however, that bacon cannot be cured except by the aid of mechanical appliances. On the contrary, a little judgment only will enable the farmer or cotter to cure whatever bacon or hams are required, and at a trifling cost. It is necessary to obtain moderately low temperatures to commence with, and then the great point in the small way is to select a situation for the curing process beneath the ordinary ground level, so that coolness and constancy in temperature be obtained.

The conditions to be observed in curing have been set down as—

(1) A uniform coolness in the cellar.

(2) A uniform strength in the pumping pickle, and

(3) Scrupulous cleanliness in all operations.

In an ordinary house or shop cellar it is possible to maintain a temperature of 55° Fahr. throughout the summer by care-

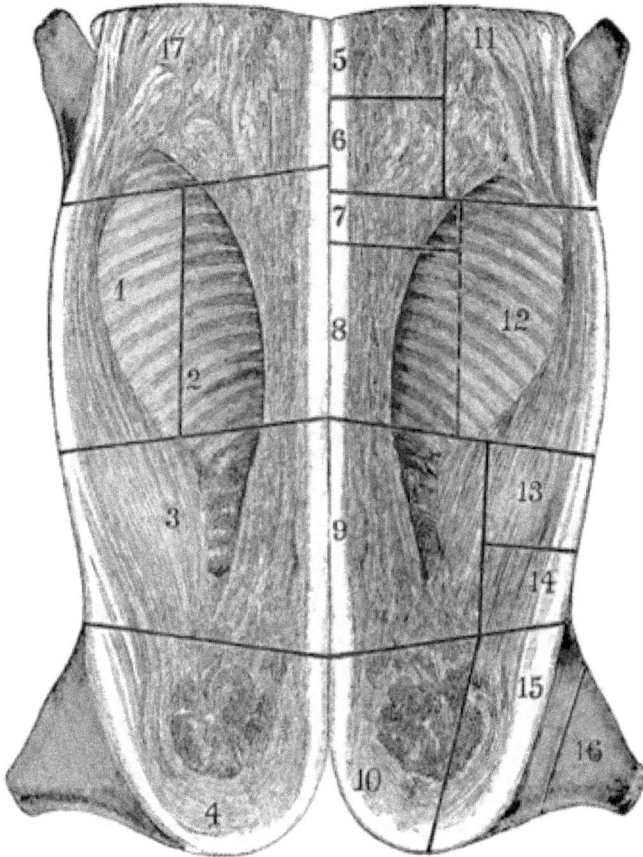

MODEL SIDES OF BACON.

[By permission of the British Dairy Farmers Association.]

| | Present Prices. per lb. | | Present Prices. per lb. |
|---|---|---|---|
| 1. Streaky Quarter ... ... | 11d. | 10. Fillet ... ... ... | 10d. |
| 2. Rib Quarter ... ... ... | 11d. | 11. Shoulder ... ... ... | 6d. |
| 3. Middle Quarter ... ... | 8½d. | 12. Prime Streaky ... ... | 11d. |
| 4. Ham Quarter ... ... | 8½d. | 13. Thin Streaky ... ... | 8½d. |
| 5. End of Neck ... ... | 7½d. | 14. Flank ... ... ... | 6½d. |
| 6. Middle of Neck ... ... | 8½d. | 15. Middle of Gammon ... | 11d. |
| 7. Thick Back and Sides ... | 10d. | 16. Knuckle of Gammon ... | 7d. |
| 8. Prime Back and Ribs ... | 11d. | 17. Fore End ... ... ... | 6d. |
| 9. Loin ... ... ... ... | 10d. | | |

fully closing up all openings, except a very small one for ventilation, and placing some ice in an elevated position.

Many cellars have been constructed on a plan suggested by the writer in his " Manual of the Pork Trade " in connection with shops, and for all purposes the same cellar can be constructed in farms or houses.

The description is as follows :—

## CELLAR.

" Wherever possible the entrance should be outside of the shop, and entirely cut off from the work room or cooking room, so as to ensure the air being fresh and free from heat from these places.  The best plan, where room is of value, is to have the cellar directly under the shop, the rafters supporting the floor being covered over so as to form a roof, either with ordinary matchboarding or sack-cloth.  It will be found most convenient to arrange round the four walls—supposing the space is square or nearly so—elevated platforms of stone work, solid and cemented perfectly smooth on the top.  The size of these should be 2 ft. high by about 3 ft. wide, and they should all be built on a gentle slope, so that the brine formed by the salt running from hams or bacon may easily run into a common channel.  At one corner it will be found necessary to have a cistern for receiving this brine, the size of which will in every case be determined by the amount of space devoted to the platform.  The brine so collected will be found very useful for replenishing the pickling vats.  In the centre of the floor may be erected the pickling vats, and these are always best made of polished slate, bolted and cemented together, or of hard Caithness flag-stones treated in the same manner.  A very useful size of vat is about 4 ft. deep by 3 ft. 6 in. square.  If it be possible to give separate space to the brine vats, say outside the cellar in underground premises, the centre of the cellar floor should be made into a large square platform, same height as the side tables, and of a size to allow of a clear footway between it and the side platforms all round."

In addition to the foregoing description, it may be added that in practice it has been shown that it is desirable to have a small aperture, with shutter ventilator inserted, so as to permit of the escape of any foul air which may accumulate near the ceiling. Should the situation of the cellar not be in a sufficiently cool place, that is to say, if a temperature of at least 55° Fahr. cannot be maintained, it will be necessary to use ice. At 55° Fahr. curing may be conducted with some little loss, but at a lower temperature (45° Fahr.) the process may be conducted with perfect immunity from taint. The position for the ice receptacle is near the roof of the cellar, and it may be made of a large tray lined with lead, upon which is set a crate made of iron rods for containing the ice. This crate may then be surrounded with louvres, so that the cold air produced by the ice will descend; the warmer air will ascend and travel over the ice surface and so be cooled and descend again, thus producing a gentle current. A temperature of 50° Fahr. will by this means be obtained. The cubical area occupied by the ice in producing this result must not be less than one-sixth of the total cubical contents of the cellar.

### CHILLING.

At all times it is desirable to chill meat before it is put in the curing cellar, and this can be done on the small scale by a "refrigerator." This appliance consists of thoroughly insulated walls and an ice receptacle, one-third of the total cubical area. The process of chilling is more carried out in large businesses than in small, and is generally performed by a refrigerating machine. On the farm it may be neglected, but in larger concerns it becomes necessary, in consequence of the meat of freshly slaughtered animals which have travelled containing a larger proportion of animal heat than would those which are thoroughly rested on the farm. Where refrigerating machinery is available, the chill room should be cooled down to 38° Fahr., and the meat should not be taken out until it indicates on the meat thermometer 40° Fahr.

## DESCRIPTION OF SLAUGHTERING AND COOLING PROCESSES.

The hogs are first of all rested in sties adjacent to the factory and are then driven in singly to the slaughtering pen. The attendant passes round one of the hind legs, below the knee joint, a slip hook which is attached to a chain, and which, in its turn, is attached to the drum of a windlass worked by either hand or power. The hogs are hoisted at once to an overhead bar, where they are slaughtered instantly by means of a sticking knife being inserted quickly into the throat, severing the jugular vein, and driven home in the direction of the heart. In the suspended position it is obvious that all the blood will flow out rapidly; so much so indeed is this the case, that movement indicating life ceases in less than one minute. The hogs are then pushed along the bar and dropped upon a table, where the hook is removed from the leg, and they are put into a scalding vat where there is a supply of water at about 175° to 180° Fahr. At this temperature the hair is loosened, and may be easily scraped off by means of either hand, bell scrapers, or the scraping machine. Hand scrapers are generally used in the United Kingdom. The pigs, when the hair comes away easily in the hand, are sufficiently scalded, and are then raised to a table by means of a "cradle" or lever elevator—a very convenient and simple arrangement pretty generally used. They are then scraped, attached by the four feet to chains, and swung into the horizontal singeing furnace, or they are swung on to a bar by the lower jaw—a "gob-hook" being used—and drawn up to a vertical singeing stack. The latter process is preferable to the former, as it saves the feet, and is besides much more economical in fuel. The singed carcases are then dropped into a cold water bath and allowed to cool for a brief space of a few minutes, and then hoisted by

the sinews of the hind legs to a bar, in which position they
are scraped with thin steel flat scrapers. They are then disembowelled, and the various portions of the viscera are
sorted, the carcase meanwhile passing onwards along the
track bar and cleansed by cold water. The backs are then
cut down and the vertebral column separated, but not cut out
entirely; the head also is cut away, being left suspended by
a skin only. The flake or kidney fat is then taken out and
placed on top of the pig, and the whole is then weighed—
minus the viscera. The weight of the hogs is reckoned on
the carcase, freed from all internal portions except the flake
or kidney fat, and this weight includes the head, the whole
being subject to 2 lbs. per hog reduction for shrinkage.
When the hogs have been weighed they are run along bars,
and the vertebral column or back bones are then cut out, and
the hogs divided into sides. The head is also cut off altogether. The sides are then allowed to hang by the sinew of
the hind leg for about six hours—until the excess of animal
heat is dissipated; they are then pushed along the bar into
the chill room, which should be at a temperature of 38°
Fahr., and are kept there until they indicate on a meat
thermometer when inserted into the gammon end 40° Fahr.
This means usually about twelve hours' chilling. At this
temperature they will be perfectly rigid and the fat hard to
the touch. They are then cut down and trimmed, the blade
bone is taken out and the feet cut off, and are then dropped
into a cellar or carried there according to the relative positions of the chill room and cellar; they are then ready to be
cured as sides.

If hams and middles are wanted, a slight modification of
the foregoing is necessary. The process is the same up to,
and so far as, the chilling is concerned, but the hams are then
separated, and the various other portions, such as middles,
shoulders, foreparts, &c., are cut out.

### Various Kinds of Meats Cured.

The principal "cuts of bacon and hams" as cured in
England are :—

Wiltshire sides.

Cumberland sides.

London cut sides.

Short clear middles.

Long fat backs.

Shoulders.

Irish hams.

York hams.

Irish sides.

Birmingham sides.

Stafford sides.

Long rib middles.

Short fat backs.

Cumberland hams.

Wiltshire hams.

There are many other trade descriptions of various "cuts" and parts of minor importance, but they are hardly of interest here. The feature of all of these is, that they do not differ in the cure, but simply in the manner of cutting. The "cut" of a ham or side of bacon is regulated by local custom.

## PROPORTIONATE PARTS OF A PIG OR HOG.

The pigs to be slaughtered should be of a size suited to the particular business for which they are intended. As a rule the size of pig preferred for all purposes is from 9 to 10 "score," or about 180 lbs. to 200 lbs. in weight. This gives a fairly sized side, about 72 lbs., without an undue proportion of fat, or a ham about 10 lbs. The actual weighing of a pig weighing 9 "score," 14 lbs. (194 lbs.) is given here as of interest in showing the turnout of a typical animal. The pig was cut up to be cured as "Wiltshire sides."

### WEIGHING OF A PIG OF TOTAL DEAD WEIGHT (WITHOUT OFFAL) OF 194 LBS.

| | lbs. | ozs. |
|---|---|---|
| Bones from back (chine) ... ... ... | 5 | 12 |
| Blade bone ... ... ... ... ... | 2 | 4 |
| Steaks ... ... ... ... ... ... | 2 | 4 |
| Cuttings... ... ... ... ... ... | 2 | 0 |
| Kidneys... ... ... ... ... ... | 0 | 6 |
| Flake fat ... ... ... ... ... | 7 | 0 |
| Fat (intestinal, &c.) ... ... ... ... | 2 | 4 |
| Feet ... ... ... ... ... ... | 4 | 0 |
| Blades ... ... ... ... ... ... | 0 | 8 |
| Skull (upper part of head) ... ... ... | 14 | 8 |
| Chap (lower jaw) ... ... ... ... | 2 | 8 |
| Two sides ... ... ... ... ... | 150 | 0 |
| Total... ... ... | 193 | 6 |

It is therefore to be inferred that heavier hams or sides can only be had from larger animals, and *vice versâ*. By long experience, however, curers have arrived at the standard of from 9 to 10 "score" as supplying the largest demand.

In Cumberland very heavy pigs are cultivated for the manufacture of "Cumberland" hams, and as a consequence the sides are very fat and only saleable in certain districts.

### BACON CURING BY THE MOST MODERN PROCESS.

When the hogs are slaughtered they should be hung up in an open space or hanging house for a night or a day, or in summer for at least six hours, and should then be placed in the chill room. A thermometer should be used to ascertain the temperature of the meat while in the chill room, and the carcases should never leave there for the curing cellar until they show a temperature of 40° Fahr. This temperature is attainable by keeping the chill-room down to 38° Fahr., dry, cold air.

When this temperature has been obtained, remove the meat from the chill room and trim and dress it in sides, quarter sides, hams, middles, or whatever kinds of meat it is proposed to cure. When sides of the " Wiltshire " pattern are wanted, the following will be the process.

First of all make a pickle from the following recipe:—

|  | lbs. |
|---|---|
| Salt ... ... ... ... ... ... | 50 |
| Finest granulated saltpetre... ... ... | 5 |
| Pure cane sugar ... ... ... ... | 5 |
| Douglas's bacon-curing antiseptic... ... | 5 |

To this add 20 gallons pure water, and if there is any scum, remove it, and should that not be sufficient to clear the liquid, it must be boiled and the scum removed as it rises to the surface. The pickle thus made should indicate on Douglas's salinometer or pickle tester, 95°—should this density not be obtained, pure salt must be added until it is. Now lay the sides on a bench or pumping table, and insert the needle

of the pickle pump into all the fleshy parts of the meat
—the pump should be worked at a pressure of about 20 lbs. to
the square inch. When the meat has been pumped with this
pickle lay it down in a light bed of salt. Now have prepared
a mixture of equal quantities of fine granulated saltpetre and
Douglas's bacon-curing antiseptic, and when the sides are
lying in the salt, sprinkle this mixture over the cut surface
lightly. Immediately afterwards cover the meat with fine
salt. When "stacking" has to be done, always make a space
between by means of three pieces of beech or ash wood, so as
to allow for free circulation of air. Allow the sides to remain
in salt from nine to fourteen days (according to the size and
mildness of the meat required). At the end of that time the
bacon is practically cured, but it is customary to turn the
sides face downwards for a few days, according to convenience,
so as to allow them to drain. If the space of the curing cellar
is valuable, this process of draining can go on in an outer cellar
where the temperature may be a little higher than the cellar
itself. When the sides have sufficiently drained, wash them in
cold water, wipe them with a cloth, and if it is proposed to sell
them as "green bacon," they might be either sent out at once
or dried in the drying room and sent out then as "pale dried;"
or if it is proposed to send them out as "smoked bacon," they
should then be placed in the smoke stoves (after having been
dusted over with fine ground Canadian pea meal) for a period
of not less than three days.

The pocket holes are the weakest part of the side, and can
only be maintained sweet by dusting into them some pure
powdered antiseptic mixed with a small quantity of ground
husks.

The smoke stoves are best made of a square shape, and
should have proper ventilation in the shape of louvres at the
top, which can be closed or opened at will by means of a chain
from the ground level. A forced draught by means of a fan,
drawing the smoke through tubes of six to eight inches in
diameter, has been proved to be very successful, especially in

heavy weather.  The best material for smoking is oak sawdust.
Any hard wood sawdust may be employed, and wheat straw
may be added if sawdust is scarce.

## HAM CURING BY THE MOST MODERN PROCESS.

To make a complete success of curing hams they should
never be taken out of the chill room until they indicate by
the meat thermometer 40° Fahr.  The meat thermometer is
simply pressed into the fleshy parts of the ham and allowed
to remain for a minute or two, when the thermometer can be
read off.  This temperature is attainable by keeping the chill
room down to 38° Fahr., dry, cold air.  When the hams are
in this condition throw them into a pickle of the following
constitution :—

|  |  | lbs. |
|---|---|---|
| Salt ...        ...        ...        ...        ...        ...        ... | | 50 |
| Finest granulated saltpetre        ...        ...        ... | | 5 |
| Pure cane sugar        ...        ...        ...        ...        ... | | 5 |
| Douglas's bacon-curing antiseptic...        ...        ... | | 5 |

To this add 20 gallons pure water, and if there is any
scum remove it, and should that not be sufficient to clear the
liquid, it must be boiled, and the scum removed as it rises to
the surface.

This pickle will be more liable to go wrong than any other
in the curing cellar, owing to the fact that the blood from the
blood vein will be drawn out ; indeed, this is the intention of
putting the hams into pickle to start with, so that the blood
may be drawn from the blood vein and thus prevent any
possibility of taint.

Allow the hams to remain in this pickle for twelve hours
and then remove and pump with a clear pickle of the same
constitution.  It must be noted here that on no account must
the pickle in which the hams are thrown for twelve hours be
used for pumping.

To restore this pickle, which will contain a quantity of
blood, it will be necessary to boil it.  To altogether dispense
with any risk, however, it is generally advisable to make a
fresh pickle when the original becomes too much contaminated.

When the hams have been pumped with fresh pickle in all the fleshy parts, lay them in beds of salt in such a position that the cut surface or fleshy parts will be level, and the shank pointing downwards. Lay the hams side by side as close together as they will go, and as each is laid down, sprinkle over the cut surface an equal mixture of saltpetre and Douglas's bacon-curing antiseptic. This mixture should always be made beforehand. Put on sufficient of the mixture to lightly cover the whole surface, and press with the finger an extra quantity into the opening of the blood-vein; on the top of this sprinkle heavily fine salt, and allow the hams to remain thus for three days. At the end of that time, take them up one by one and squeeze out all that may remain of the coloured liquid from the blood-vein, wiping the same clean from the cut surface. Immediately then lay them down in beds of salt as before, and cover them up with fine salt. The smallest ham should now remain at least fifteen days in this position, and it is customary where hams are to be kept for a period to allow them to remain twenty-one days. After twenty-one days, the length of time they have to remain is according to the rule that it requires one day to cure 1 lb. meat, so that heavier hams will require one day for every lb. above 21 lbs. At the end of this period take the hams up, wash them in cold water, wipe them clean, and if they are to be sold as "pale dried," dry them in hot air rooms at a temperature not exceeding 80° Fahr. until they are quite dry.

In order to procure the excessive paleness which is characteristic of some hams, they are dipped in scalding hot water, and then allowed to dry before sending out.

If the hams are to be smoked they are taken up from the cellar, washed, and immediately hung in smoke stoves, where they are allowed to hang not less than three days.

To produce the glossy appearance familiar in some smoked hams, a little vaseline is rubbed over the skin.

# INDEX.

12

180 INDEX.

www.ingramcontent.com/pod-product-compliance
Lightning Source LLC
Chambersburg PA
CBHW021703210326
41599CB00013B/1500